The Institution of **StructuralEngineers**

SECOND EDITION

Structural robustness and disproportionate collapse in buildings

Authors (2nd edition: 2023–)

R Haynes   MA(Cantab) MEng CEng FICE (Quadrant Building Control, a Socotec Company) *Lead*

J Burvill   MEng MSt CEng MIStructE (Laing O'Rourke)
E Halliwell   MA(Cantab) MEng CEng MICE MIStructE (MPA The Concrete Centre)
S Hanlon   BEng MSc CEng MIStructE (SPHstructures)
G Lewis   MEng CEng MICE MIStructE (Milner Associates)
M Milner   MSc CEng MICE MIStructE (Milner Associates)
E O'Loughlin   BEng MSc CEng MIFireE PMSFPE (Arup)
B Price   MEng ACGI CEng MICE (Tamarind Engineering Ltd)
J J Roberts   BSc(Eng) PhD CEng FIStructE FICE FIMS FCMI FICT Hon DSc
A Soane   BSc PhD CEng FICE FIStructE HonIFE (CROSS-UK)
A G J Way   MEng CEng MICE (The Steel Construction Institute)

Reviewers (2nd edition: 2023–)

Patrick Hayes (IStructE Technical Director), Tony Jones, Luke Bisby, Danny Hopkin, Andrew Lawrence, Susan Lamont, Allan Mann, Katarina Lizuchova.

Publishing (2nd edition: 2023–)

L Baldwin   BA (Hons) DipPub (The Institution of Structural Engineers)
J F Koernich   (JFK Publishing Services)
R Thomas   BA (Hons) MCLIP (The Institution of Structural Engineers)

Printed by Cambridge Scholars Publishing

Published by The Institution of Structural Engineers
International HQ, 47–58 Bastwick Street, London EC1V 3PS, United Kingdom
T: +44(0)20 7235 4535
E: mail@istructe.org
W: www.istructe.org

First published (version 1.0) November 2023
978-1-906335-66-3 (print)
978-1-906335-67-0 (pdf)

Authors (1st edition: 2010–2023)

A P Mann FREng BSc(Eng) PhD CEng FIStructE MICE (Jacobs) Chairman
S J Alexander MA CEng FIStructE FICE MCMI (WSP Group)
J N Carpenter BSc(Eng) CEng FIStructE FICE CFIOSH (SCOSS)
J P Cartz CEng FIStructE (Capita Symonds/Capita Architecture)
M Chryssanthopoulos BSc MS PhD DIC CEng FIStructE FICE (University of Surrey)
G T Harding OBE DIC CEng FIStructE MICE (Consultant)
A E K Jones BEng(Hons) PhD CEng FICE (Arup)
P Kelly BSc(Eng) CEng MIStructE (Treanor Pujol Limited)
G Lewis MEng(Hons) CEng MIStructE MICE (CCB Evolution Ltd)
A Thirumoolan BSc(Hons) CEng MICE FRICS (Wandsworth Borough Council)
J N Tutt MPhil CEng FIStructE (Jenkins and Potter)
T C Cosgrove MSc DIC CEng MIStructE MIEI (SIAC Tetbury Steel)

Secretary to the task group (1st edition: 2010–2023)

B Chan BSc(Hons) AMIMechE (The Institution of Structural Engineers)

Acknowledgements

The Institution would like to thank Collaborative Reporting for Safer Structures UK (CROSS-UK) for pinpointing the appropriate CROSS reports to be referenced and signposted, and Lauren Evans-Haynes for her work in creating Figures 2.2, 2.3, 2.11, 2.12, 3.3, 3.4, 5.5, 5.6, 6.2, 7.1–7.4, 9.3, 9.6, 9.7, 10.15, 10.17–10.20, 10.23, 10.24, 11.2 and 11.6–11.10.

Permission to reproduce the following has been obtained, courtesy of these individuals/organisations:

Cover photo: Champlain Towers South, Florida (Image courtesy of NIST. All rights reserved, US Secretary of Commerce)
Figure 2.1: Alan Stanton CC BY-SA 2.0 via Wikimedia Commons
Figure 2.5: Eric Brinkhorst/Hollandse Hoogte/eyevine
Figure 2.6: Colin Jolly
Figures 2.8 and 2.9: Crown copyright — Department of Education and Science
Figure 2.13: Crown copyright — Ministry of Housing, Communities and Local Government
Figure 2.15: dpa picture alliance/Alamy Stock Photo
Figure 3.1: Philip Halling CC BY-SA 2.0 via Wikimedia Commons
Figure 3.2: Derived/adapted from BSI — BS EN 1991-1-7
Figure 3.5: Derived/adapted from IStructE *Manual for the systematic risk assessment of high-risk structures against disproportionate collapse*
Figures 5.4 and 6.1: Crown copyright — HM Government Approved Document A
Figures 5.7 and 5.8: BSI — NA to BS EN 1990
Figure 6.3: Abaca Press/Alamy Stock Photo
Figures 8.1 and 8.2: Patrick Hayes
Figure 8.3: The Steel Construction Institute
Figure 9.1: IStructE *Manual for the design of concrete building structures to Eurocode 2*
Figure 9.2: BSI — BS EN 1992-1-1
Figure 9.4: Associated Press/Alamy Stock Photo
Figure 9.5: The Concrete Society
Figures 10.3–10.5, 10.8, 10.9 and 10.11–10.14: MPA The Concrete Centre
Figure 11.1: Clem Rutter CC BY-SA 3.0 DEED via Wikimedia Commons
Figure 11.3: Brick Development Association, Aircrete and Concrete Block Association
Figure 11.5: PA Images/Alamy Stock Photo
Figure 12.1: BSI — PD 6693-1
Figures 12.7: Structural Timber Association
Figure 12.10: Tamarind Engineering Ltd
Figures 12.11 and 14.3: Arup
Figure 13.2: Paul Bell
Figure 14.2: Luke Bisby
Figures 14.4–14.6: With permission from ASCE

Table 2.1: Crown copyright — Ministry of Housing, Communities and Local Government
Tables 4.1 and 11.1: Derived/adapted from BSI — BS EN 1991-1-7
Table 4.2: Patrick Hayes
Table 5.1: Derived/adapted from BSI — BS EN 1991-1-7 and Crown copyright — HM Government Approved Document A
Table 9.1: Derived/adapted from IStructE *Manual for the design of concrete building structures to Eurocode 2*
Table 11.2: Derived/adapted from BSI — PD 6697
Table 13.1: Derived/adapted from IStructE *Appraisal of existing structures* (3rd edition)

Permission to reproduce extracts from British Standards is granted by BSI. British Standards can be obtained in PDF or hard copy formats from the BSI online shop: www.bsigroup.com/Shop

Contents

Foreword

During my university education in structural engineering 50 years ago, there was essentially no treatment of the very important subject of this book. We learnt how to design structures to the relevant codes and standards on an element-by-element basis, whether the material was structural steel, reinforced or prestressed concrete, masonry or timber. We rarely considered structures as systems. Only in courses in seismic engineering in graduate school at the University of California at Berkeley was I exposed to the concepts that are so important to robustness and resistance to progressive collapse. For many, seismic engineering was considered a subject only for 'west-coast' US structural engineers at that time.

In the decades since then, I've had opportunities to design some innovative structures, requiring deep reflection on concepts of structural reliability, and to investigate the causes of some of the most catastrophic structural failures of our time. I believe the development of my comprehension of the importance of robustness and resistance to disproportionate collapse reflects the general understanding of our international structural engineering community — for it has been, and continues to be, an evolution.

The failures at Ronan Point in 1968 and at the Skyline Towers in 1973 catalysed attention to progressive collapse in the UK and US respectively. Our knowledge and practices have advanced greatly since then but are still improving.

Early in my career I co-led an investigation of the devastating Hyatt Regency Walkways collapse in Kansas City, Missouri in 1981. At a loss of 114 lives, the Hyatt collapse remains the deadliest accidental structural failure in US history. The lesson for us all is how clearly avoidable this tragedy was, and that lesson attests to the critical importance of the subject of this book. Through attention to robustness and resistance to progressive collapse — tying together the structural elements, providing redundancy and ductility, and building alternative load paths into the structural system in the event of a local failure — we can prevent an initial event from becoming a tragic loss of human life. As this book demonstrates, robustness can often be obtained at little or no additional cost to the structure.

More recently, the disastrous fire at Grenfell Tower in London in 2017 and the devastating partial collapse at Champlain Towers South in Florida in 2021, are important reminders of the dreadful human toll of catastrophic events. (While Grenfell was not a structural failure, *per se*, it highlights some of the important concepts in building design that can prevent catastrophic events, such as the need for 'compartmentalisation', and it is transforming building regulations in the UK.)

Today, the following trends demand increased attention to robustness and resistance to progressive collapse:

- Computer programs now allow structural engineers to design individual elements of a structure to the minimal margins acceptable by codes and standards, reducing the degree of 'overstrength' common when laborious hand methods encouraged us to design with a blunter pencil
- Our essential need to design more materially-efficient structures to reduce embodied carbon, requires us to optimise the use of materials in providing structural reliability. To maintain structural reliability, we need more systems-thinking in our design methods
- Climate change is increasing the severity and frequency of some of the natural hazards our structures must resist
- Resistance to intentional events (e.g., terrorist) is increasingly important

The primary author of this second edition, Ruth Haynes, is particularly qualified to provide a contemporary update to this book. She is not only a very talented structural engineer herself; she is a leading educator of structural engineers. She is an expert in the technical nuances of structural behaviour under extreme circumstances and can readily address building regulations and control. Ruth has lectured on such diverse subjects as the Party Wall Act, building collapse and cyclone shelter design.

Structural robustness covers the spectrum; from theoretical concepts to current practices, codes and standards, and accommodates the evolutionary nature of the subject.

The state of practice is not purely prescriptive, and the book describes the variety of approaches that may be employed in the general categories of:

- Robustness and redundancy through 'tying' the structure together
- Hypothesising the removal of notional elements
- The design of key elements

The need for fundamental understanding of structural behaviour and application of engineering judgement is emphasised:

"Structural engineering should never become a totally mathematical exercise; the real strength of structures is a function of their theoretical design, the quality of detailing and the quality of construction."

Particularly unusual for treatment of this subject is the frequent illustration of concepts through the failure and near-miss case studies of the *Collaborative Reporting for Safer Structures* (CROSS) programme. This is especially relevant to the book's subject as it connects the theoretical aspects of robustness and progressive collapse to real-world evidence of the origins of structural failures.

This second edition provides key additions and enhancements to the first edition. It follows the latest Eurocodes and integrates the new UK Building Safety Act, enhancing both through case studies. There are new chapters on risk, classification of existing buildings, lightweight steel frames, and alterations to existing buildings. The last subject responds directly to the need to extend the life of existing buildings as a means of limiting embodied carbon in our built environment.

This second edition of *Structural robustness and disproportionate collapse in buildings* is an important resource in the evolution of our practice in avoiding catastrophic failure. It is for structural engineers of all experience levels, students just starting out in the field, and anyone with an interest in the built environment. While it is British Standard- and Eurocode-centric, it is nevertheless a valuable resource for the entire international community.

Glenn R. Bell FIStructE, FICE, F.SEI, Dist.M.ASCE
Acton, Massachusetts

1 Introduction

At just after midnight on 14 June 2017, a fire broke out in the 24-storey Grenfell Tower block of flats in North Kensington, West London. Just four hours later, Elpidio Bonifacio left the building. He was the last resident to leave alive.

72 people died and more than 70 others were injured. It was the deadliest fire in a residential building in the UK and a completely unacceptable tragedy.

The prime minister at the time, Theresa May, announced a public enquiry into the fire and in December 2017, Dame Judith Hackitt unveiled the conclusion of her interim report, describing the system of building regulation as not fit for purpose[1].

The Grenfell Tower fire brought into sharp focus the deficiencies within the construction industry, and the publication of the Building Safety Act[2] in 2022 reinforced the legal responsibilities of structural engineers to protect the public.

One of the requirements of the Act is that a safety case will be required for higher-risk/'in scope' buildings, to demonstrate that a building is safe from structural collapse. This is an important time to remind engineers that designing for robustness is not an 'add-on' at the end of a design, or just a token set of calculations, but a hugely important aspect to be thought through carefully from the start of a project. It is therefore appropriate that the IStructE *Practical guide to structural robustness and disproportionate collapse in buildings*[3] has been revised to highlight the responsibilities of engineers, and to help with production of a suitable design.

BS EN 1991-1-7[4] defines robustness as:

"The ability of a structure to withstand events like fire, explosions, impact or the consequences of human error, without being damaged to an extent disproportionate to the original cause."

The terms 'robust building' and 'disproportionate collapse' are used interchangeably in building design, but the general understanding is that if a building is robust, it will not collapse disproportionately in the event of an accidental or extreme event.

The purpose of this *Guidance* is to enable practising structural engineers to feel confident in designing a robust building, such that design against disproportionate collapse becomes an integral part of the structural design process.

Structural engineers design with many different structural forms and materials, but apply very similar principles to all buildings; that is to transfer vertical and horizontal loads safely to the ground.

Engineers are less confident when it comes to designing to prevent disproportionate collapse because application of the rules is different for each material used, and mixed design concepts are involved; they include complex formulae that are applied in an empirical way. Risk is considered in a general sense by the engineer but is often not quantified.

This *Guidance* includes case studies for each building material, with the intention that once a structural engineer is confident with the design rules outlined in the Eurocodes, sound judgements can be made when creating the strategy for designing a robust building, and the more nebulous concepts of robustness can be dealt with.

Also highlighted are the differences between materials in designing for robustness, and how some of these differences are historic and related to the innate properties of the material. Concrete, steel, timber, masonry, precast concrete buildings, modern methods of construction and also alterations to existing buildings are covered.

Designing structural alterations to existing buildings is one of the most difficult challenges for structural engineers, because the original buildings to be adapted would not necessarily have been designed with robustness in mind. However, with the Building Safety Regulator insisting on a structural and fire safety case assessment for existing buildings, it is key that engineers understand how to assess the robustness of these structures.

The *Guidance* is set out so that general concepts of robustness and structural design to prevent disproportionate collapse are covered in Chapters 2–6, and the material chapters, with case studies, are in Chapters 7–12. Alterations to existing buildings are discussed in Chapter 13, and Chapter 14 offers information on structural robustness in fire both for designing new structures and adapting existing buildings. The material chapters are relatively stand-alone, however Chapters 2–6 and Chapters 13–14 are a vital read, regardless of the specific material involved.

Finally, the word 'accident' is used frequently because the Eurocodes use 'Accident/Accidental Action, A', to describe an event that may impact on a building's structure and within design load combinations. In the context of this *Guidance*, 'accident' therefore means an extreme loading event, rather than something quite trivial. The words 'loads' and 'actions' are used interchangeably according to English language useage, but also to reflect the terminology used in the Eurocodes.

2 Concepts of robustness

2.1 Introduction

A robust building is a building that does not respond disproportionately to an event that may arise during its design life.

This event may be an error in construction, an accidental or extreme event during the life of the building, or a detail or design principle that has not been sufficiently considered by the structural engineer. The engineer must design the building so that if one of these problems occurs and there is a collapse, that collapse is limited to a small proportion of the building.

The construction industry has become more sophisticated and complex, with more design responsibilities passed on to the contractor and value engineering being common. This leads to a project being designed by multiple structural engineers and consequently an increased likelihood of design errors or omissions.

The public expects to be safe in buildings and relies on the technical expertise and integrity of the design and construction team. Therefore, it is imperative that the design and construction team communicate effectively.

2.2 Hybrid structures

Primary and secondary structures of a building are commonly designed by different consultants and contractors.

For example, the foundation design, if piled, is likely to be carried out by the piling contractor using loading provided by the structural engineer who designed the superstructure. The floors, if precast, will be designed by the precast concrete manufacturer, and the roof, if trussed rafter, will be designed by the trussed rafter manufacturer. If the building is in steel, the steel connections will be designed by the steel fabricator to forces provided by the structural engineer.

At every intersection of different materials and designers, the approach to disproportionate collapse needs to be consistent. If it has been decided that vertical and horizontal ties are to be provided (a prescriptive approach to robustness), they need to be provided across every structural interface. Failure investigations often pinpoint interfaces as areas of weakness.

Another complication is that it is often a structural engineer from a firm of consulting engineers who acts as the main structural designer for the project, with overall responsibility for the whole building. However, it is the manufacturers of piles or precast floors who understand their products in more detail. It is therefore important that all the structural designers know who is responsible for the robustness strategy for the building, and that they support and assist the design engineer responsible for overall building safety.

The robustness of a structure must be looked at as a whole.

In all structures, and particularly in hybrid structures, there must be one designer who takes primary responsibility for assuring stability and robustness of the whole structure, and who defines and documents the strategy. This principle does not, however, remove the need for other individual designers to take responsibility for the robust design of the various elements that make up the whole.

Equally, the same point applies during construction. Risks of failure may be highest in the partially-completed state, so one designer should have overall responsibility for assuring robustness/stability during the construction phase.

✝ *CROSS Safety Alert: Structural stability/integrity of steel frame buildings[5]*
A significant alert on the total collapse of the City Gates Church building in Ilford, a steel-framed building under construction, due to issues of design, detailing and construction. The building collapsed on 31st January 2012. The most likely failure scenario was that some areas were weakened by misalignment of welded joints, undersized welds, and double-drilled gusset plates. Once loaded with building materials, this may have resulted, initially, in slow plastic yielding followed by brittle fracture of some of the thicker sections.

✝ *CROSS Safety Report: Disproportionate collapse assessment of large panel system buildings[6]*
This report drives home the argument that sometimes, to achieve robustness in all cases, cooperation between disciplines is necessary to achieve that holistic perspective. In this case, fire engineers and structural engineers work together to ensure that a large panel system building does not disproportionately collapse due to fire.

✝ *CROSS Safety Report: Risks from off-site manufacture and hybrid construction[7]*
This report is about a 'near miss' involving concrete construction in which precast and *in situ* concrete were used in combination.

✝ *CROSS Safety Alert: Safety issues associated with balconies[8]*
This is a report about multiple issues associated with balcony design, including procurement, where it is common for several designers to be involved.

2.3 Structural form

If a building can transfer horizontal and vertical loading to the ground in a straightforward and direct way, it is more likely to be a robust building.

In a traditional cellular masonry building, with walls aligning vertically on each level, it is relatively easy to understand what parts of the building may collapse if a wall is blown out by an explosion. It is the same with steel or concrete buildings where the column layout is the same on each level; if a column were to be lost, it is not difficult to consider what might occur at higher levels of the building.

However, if the building has a more complex structural form, with transfer structures or cantilevered floors, it is more difficult to determine what would occur if one of these structural elements were to be lost, although easily understood that collapse would be more extensive.

The load path of a building should be defined at the start of the design process and, with hybrid structures, conveyed to all of the design team. It is not good practice to start off with a design with inherent weaknesses in its structural form or an indirect load path. Many structures do have indirect load paths and if this is the case, particular care in the design is needed.

Redundancy in a structure is also key to a robust building. If a structure is statically determinate, it can revert to a mechanism if one part fails. Figure 2.1 shows a bus station with single columns supporting a roof structure. If the single column were to fail, the roof would fall.

Figure 2.1: Tottenham Hale Bus Station, single column supporting the roof

Another example of this is a pinned portal structure. If a column were to be lost, the portal frame becomes a mechanism and cannot remain standing (Figure 2.2).

Figure 2.2: Portal frame structure with pinned base

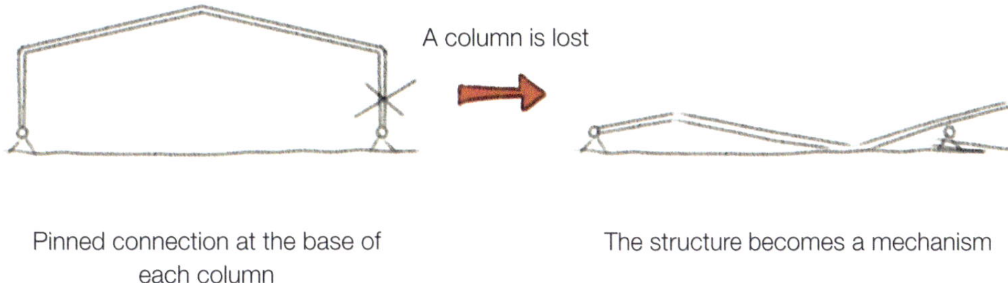

Pinned connection at the base of
each column

The structure becomes a mechanism

If the portal frame column bases were to have a four-bolt connection to the foundation, the remaining column could stay upright, assuming that the foundation was sufficiently robust, and there is a possibility that the combination of the portal rafter hanging from the remaining column and the purlins makes failure less catastrophic (Figure 2.3).

Figure 2.3: Portal frame structure with four-bolt connection at base

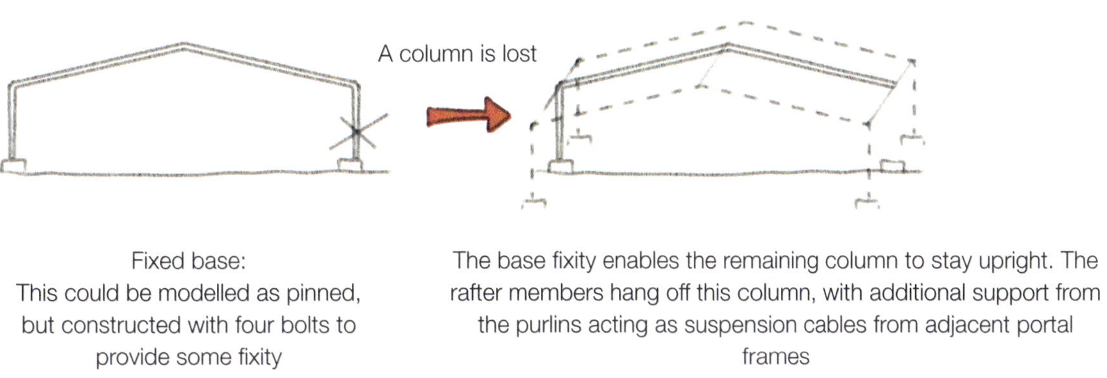

Fixed base:
This could be modelled as pinned,
but constructed with four bolts to
provide some fixity

The base fixity enables the remaining column to stay upright. The
rafter members hang off this column, with additional support from
the purlins acting as suspension cables from adjacent portal
frames

It is also generally accepted that the more ductile a structural form is, the more robust it is. Ductility desensitises the entire structure, or its individual components, to damage from uncertainties inherent within loading and construction, and to variations from predicted stress levels.

Ductility is relevant at component level and in connection design. It is relevant in static loading but also key for buildings subject to dynamic loads such as seismic loading (usually not required in the UK for standard buildings) or cyclical loading. In steel structures, the standard connection details have some ductility and in reinforced concrete, detailing rules also provide a level of ductility.

CROSS Safety Report: Concern about design principles for tall buildings[9]
This report is about very tall and complex reinforced concrete structures that incorporate combined transfer systems, some of which may be supported on corner columns.

CROSS Safety Report: Concerns over robustness of some 20 year old buildings[10]
This report is about a masonry building constructed 20 years earlier. The structure consisted of masonry support walls and precast floors with thin walls, small support piers, and short or no buttressing returns.

2.4 Resistance to horizontal loads

The structure of a building should provide a clear path to transfer horizontal loads to the ground, but there are some structures where it is not obvious what the horizontal loads are.

The majority of buildings will experience wind loading, but there are other horizontal forces that may be induced due to an out-of-plumb column or slight building sway, eccentric loads or vibration.

In some situations, such as an internal independent mezzanine within a building with no dominant openings, wind loading will be minimal (Figure 2.4). However, it is expected that engineers design for a minimum horizontal loading to account for issues such as tolerance or sway. Section 5.3 further explores the subject of horizontal loading and notional inclination.

Figure 2.4: Sheltered internal structure with heavy floor loading yet no defined explicit horizontal load to assure stability

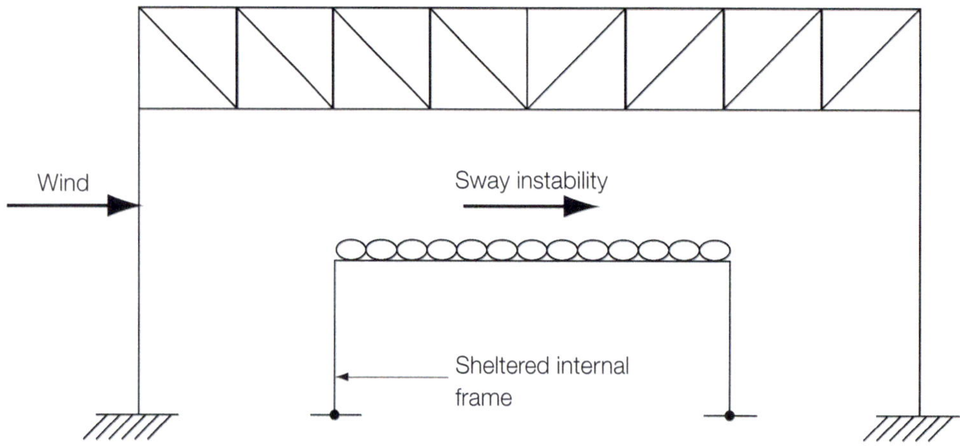

Robustness requires the ability of a building to resist loads during all stages of construction.

Structural elements may be subject to more onerous loads during construction or have less support than once the building is complete, and this needs to be considered during design of temporary works and sequencing of construction.

During construction, many elements that are eventually sheltered within a completed building are temporarily exposed to wind loads, and there are occurrences of walls blowing over before a building is completed. Consequently, free-standing walls are typically propped during construction and before they are either enclosed within a building or tied into a floor or roof. BS EN 1991-1-6[11], which covers applied forces during construction of a building, also includes a nominal horizontal force (Fhn).

There have also been instances of tall bridge girders toppling over before the bridge deck has been added. Therefore, it is important to consider the forces that may apply during construction as well as once a structure is completed, and that a temporary stability system is designed. The design of temporary works to enable a structure to be safely built are as significant as the design of the permanent structure. The Temporary Works Forum (TWF)[12] is a useful resource, and the structural engineer should also refer to The Construction (Design and Management) Regulations 2015 (CDM)[13], that require all risks to be managed from the start to the end of a project and address the buildability of the permanent works.

CROSS Safety Report: Stability of tenants' mezzanine[14]
This report explains the issues of lateral resistance to horizontal loads of mezzanine structures.

The Construction (Design and Management) Regulations 2015[13]

Health and safety at work, etc. act 1974[15]

Temporary works toolkit[16]

2.5 Components, connection design and ductility

Much of modern structural engineering relies on using components that are rated at maximum efficiency, with such efficiency gained by all elements contributing structurally in some manner.

A good example is the steel pitched portal frame with haunched eaves, where the lower eaves compression flange is restrained out-of-plane by proprietary braces back to lightweight cold-rolled purlins, which are themselves restrained by the roof covering and in-plane sag rods.

Such systems are economical and widespread, but it is important to remember that the different structural components are interdependent for their structural performance. Therefore, care must be taken not to remove some of these apparently minor components in case that removal has more widespread consequences, potentially setting off a chain reaction of collapse. The FC Twente stadium roof in the Netherlands is an example of this kind of collapse (Figure 2.5)[17].

Figure 2.5: FC Twente stadium roof collapse

In the absence of buckling, beams are robust under vertical load if they have strong connections and if they exhibit ductility; this is assured by codified rules such as controlling reinforcement percentages in concrete members or the elimination of local flange buckling in steel.

Where there is a possibility that failure of a beam or column is caused by buckling, robustness requires that lateral stability is not dependent on a few insubstantial bracings or restraining members, which might raise the danger of gross failure linked to loss of minor members.

Similarly, it is not good practice to skimp on beam end connection capacity. Whatever the design forces, there should be correlation between beam capacity and the connections that support it (including the need for connections to act as ties).

Alongside robustness, ductility is a sound attribute to have, and achieving it is partly a matter of design and partly a matter of detailing. Without ductility, structures would be vulnerable to abrupt failure, and engineers could not rely on design procedures such as slab yield line analysis in concrete or plastic design of steel frames.

At a simple level, ductility allows constant shear to be carried as in a real hinge, and a sound objective of steelwork detailing is to allow shear connections to deform, yet still carry normal loading even under working conditions. At a more advanced level, ductility allows moment to be carried at constant magnitude when deformation under plastic hinge conditions takes place.

The objective of standard detailing rules in steel and concrete is to ensure that the plastic hinge has sufficient capacity to permit load redistribution while undergoing further deformation. This can be quantified through the shape of the moment–curvature curve, specifically the part of the curve after maximum moment capacity is attained.

The area under the moment–curvature curve can also be considered as a measure of the ability to absorb energy by the component. Through this concept, structures can be assessed with respect to shock dynamic loading, e.g., gas explosions, blast, vehicle impact and earthquake action, since all those incidents release a finite amount of energy. Consequently, quantifying and mobilising energy absorption capability is a key constituent in robustness strategies against accidental loading.

It is unlikely that a design engineer will carry out detailed calculations on the energy that can be absorbed by a connection, and will follow the material codes on connection design and detailing.

The practice of adopting minimum sizes and slenderness ratios, minimum reinforcement and bearing, etc., assures a certain amount of construction robustness. However, the structural engineer does need to be aware of the structures where following standard detailing and design rules is not sufficient.

Self-evidently a robust structure can only be achieved if the quality of detailing and construction is commensurate with the design intent.

The ability to absorb energy is a key quality of robustness.

Connections and supporting structural elements must be able to absorb energy to allow significant rotation or deflection without fracturing.

Looking at dynamic effects in terms of applied force suggests the forces involved can be very high with correspondingly high stress, which is alarming yet misleading; a high force exists but it will be of very short duration, although may cause permanent damage.

A good illustrative example is to investigate the behaviour of a standard crash barrier post under vehicle impact. The impacting vehicle has a certain amount of energy. When the vehicle hits the post it bends, forming a plastic hinge at its base. The bulk of the energy from the vehicle is absorbed by rotation of the post (Figure 2.6).

The more a post is able to rotate without the post failing, the more energy is absorbed. However, if the post rotates too much, it will not stop the vehicle. To rotate, the post and its connection to the ground must be capable of significant plastic deformation, e.g., post and connection need to be ductile.

The example of the crash barrier post is a good comparison to the demands made on structural systems as a whole, and applies to structural elements and connections. A building that can sustain deflections and deformations rather than undergo excessive cracking and tearing if subjected to accidental loading is a robust building, and therefore a safer building.

Figure 2.6: Energy absorption via plastic bending

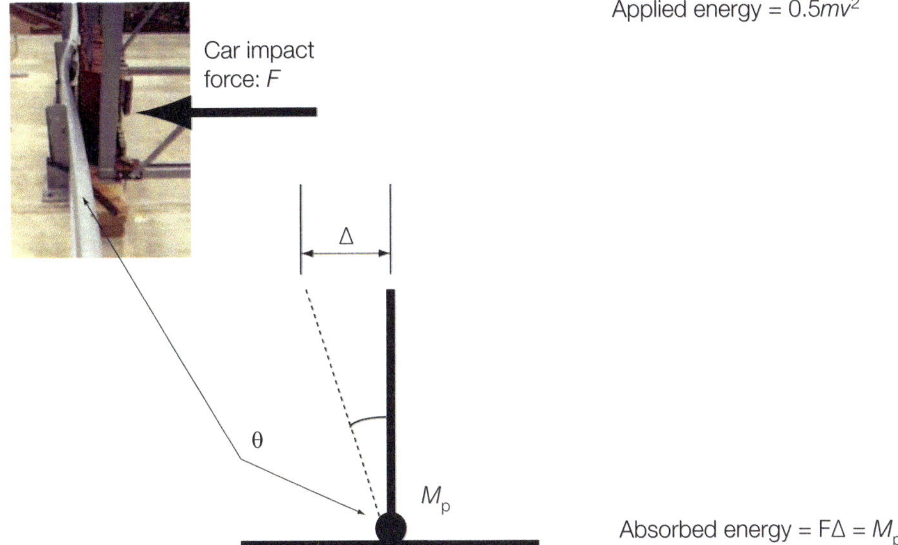

CROSS Topic Paper: FC Twente stadium roof collapse[17]

2.6 Redundancy

When failure of any one loadbearing member leads to the collapse of the entire structural system, as is the case in pin-jointed trusses, the structure has no redundancy. Statically determinate structures will revert to a mechanism if just one part fails. In this case, the structure can be thought of as a chain under tension; loss of one link implies loss of the ability to transfer load from one end to the other. This is known as a 'series' or 'weakest-link' system.

Conversely, in a parallel system, members are interconnected in such a way that load is shared between them, and failure of one member will lead to load redistribution to other members. Such a system has a degree of redundancy.

In parallel or redundant systems, further distinctions can be made depending on whether redundant members pick up loads, even when the structure experiences a low level of loading (active redundancy) or whether they participate only after a certain degree of degradation/damage has occurred (passive or fail-safe redundancy). Back-up tie cables can be used to secure objects that might fall due to fatigue or corrosion, etc.

A second important factor is the so-called 'common cause failure'. This is where members that are expected to share loads are susceptible to common underlying factors that may lead to their failure at the same time or under the same conditions (e.g., all structural elements suffering from the same manufacturing defect).

One strategy of the building regulations and codified methods of imparting robustness is to rely on a level of structural continuity and redundancy. When this is provided there may be sufficient capacity in any undamaged structure to carry loads redistributed from the damaged elements via alternative load paths.

In simple terms, redundancy means the structure has more load paths than it strictly needs. Inherently this is a good thing since damage or deficiency in any one part (joint or component) does not instantly mean total failure.

In any discussions on robustness there is a second level of redundancy or spare capacity that can be exploited — the available margin existing between what the elements are capable of carrying and what the demand actually is. Margins exist as the difference between real material strength and specified strength, because the loads on the structure during the initiating event will often be less than those used for design purposes. This is not always the case; if a structural element fails, loads higher than the design loads may be induced on adjacent structural members. The loads imposed on a structure during a major fire may also be far greater than the design loads. Codes offer reduced safety factors to allow for these facts. Load factors used with accidental loading are discussed in Section 5.8.

 *Hambly's paradox[18]*

 *Manual for the systematic risk assessment of high-risk structures against disproportionate collapse[19]*

2.7 Insensitivity to precision of design assumptions and construction accuracy

The concept of insensitivity is an attribute of robustness, and this can be seen in many ways; the errors of poor tolerance or unanticipated building movement can have a disproportionate effect. Although the execution standards NSCS[20] and NSSS[21] go some way to mitigate these issues, refined calculations are inappropriate if they ignore them, and competency and experience are important attributes for a structural engineer when considering robustness.

Structural engineering should never become a totally mathematical exercise; the real strength of structures is a function of their theoretical design, the quality of detailing and the quality of construction. There are uncertainties in each of those stages.

As a principle, designs should account for credible variations in design assumptions. This applies, for example, in reinforced concrete design where minor bar positional errors are inevitable, or in steelwork where joint shimming might be required. There have been cases of sudden failure of flat slabs in shear around column heads where bars have been badly positioned or trampled down during concreting.

It is well known that a 5mm loss of cover will have a significant deleterious effect on the durability of reinforced concrete. Good design requires that insensitivity is achieved by good detailing, by recognising the difficulties of construction and by concentrating on such matters rather than relying on structural efficiency expressed purely in numerical terms of strength.

Figure 2.7 shows a thin concrete slab that was designed as post-tensioned, with the tendon offset by 10mm. A positioning error of 10mm totally undermined the assumptions made on capacity and fatally weakened the slab.

Figure 2.7: Post-tensioned thin concrete slab

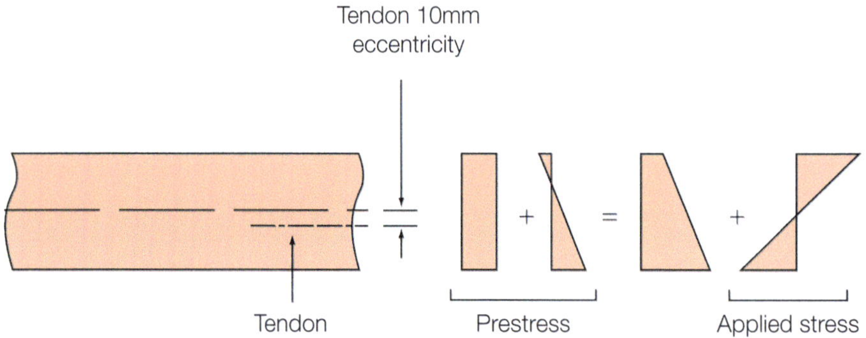

Balance of stress is very sensitive to assumed eccentricity

When designing for wind as the dominant load, care must be taken that the structure is not too sensitive. For example, it is customary for a reduced wind load to be taken during construction, but at a low speed of perhaps 10m/sec, a credible marginal increase of 1m/sec will raise the forces by 21%, as wind load is proportional to the square of wind speed.

$$(11/10)^2 = 1.21$$

To cope with such potential variance requires robustness, not more precision in the calculations.

2.8 Detailing quality affects robustness

The failure of the hall roof at the Camden School for Girls in 1973 (Figures 2.8 and 2.9)[22] was precipitated by the trivial bearing width provided for the roof beams and by corrosion of the reinforcement that was supposed to keep the beams on the support (Figure 2.10).

A robust support detail should have sufficient bearing and allow for the fact that construction tolerances may result in less bearing. Support structures should also be designed for any eccentricity inherent in a connection, whether through design, construction inaccuracies, temperature effects or shrinkage.

Figure 2.8 and Figure 2.9: Camden School for Girls roof collapse

Figure 2.10: Camden School for Girls beam bearing detail

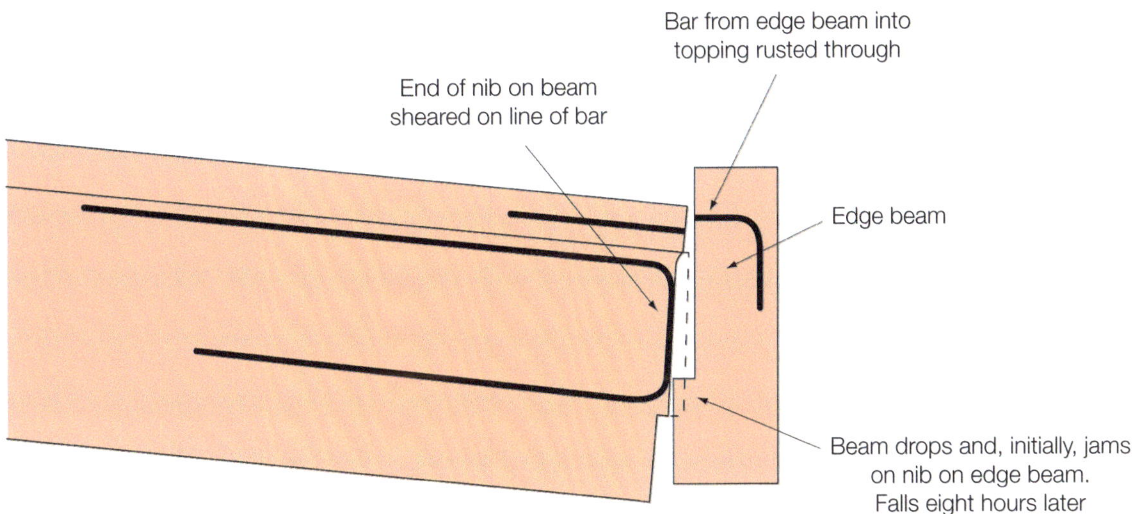

2.9 Insensitivity to building movement

When designing for robustness, the engineer needs to consider what might happen if supports move over time, e.g., due to settlement or differential thermal movement.

An example of this is a building in which precast floors span on to edge beams. If the edge beams were to bow out-of-plane, the precast floors would have less support and therefore be subject to unanticipated forces. In the worst case, they could fall between the beams.

Conversely, a supported structure can also move due to unplanned loading; if a flat roof were to sag and ponding occurred, the flat roof joists could be pulled away from the supporting walls.

A very common example is also seen in ruined buildings; the walls move outwards due to settlement and the ends of the floor joists rot due to the ingress of water. Once the intermediate floors no longer tie the walls together, there is less to prevent progressive wall movement and the roof falls in.

The requirements of safe erection also dictate that designers consider any out-of-balance loadings that might exist during flooring installation.

2.10 Secondary building structures

A common failure within buildings is that of ceiling collapse. Many buildings have suspended ceiling structures, hung from the main structure with proprietary anchors.

If one of the anchors pulls out, the adjacent hangers are subsequently overloaded. This causes an unzipping of the ceiling from the supporting structure (Figure 2.11).

Figure 2.11: Suspended ceiling hung from main structure

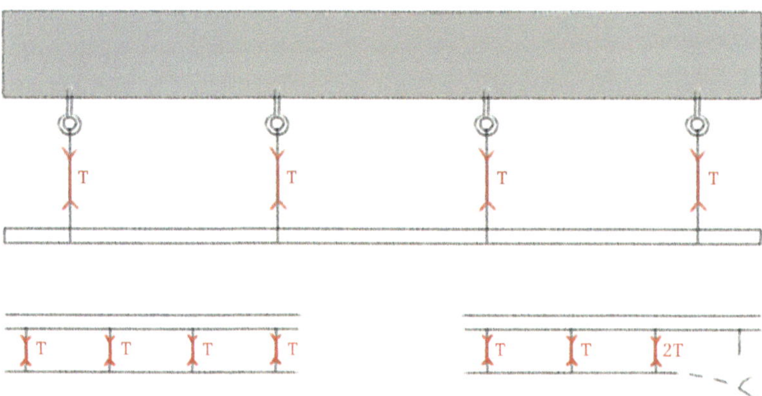

This is an example of both progressive and disproportionate collapse. Robustness demands that when possibilities for failures of this kind exist, there must be high confidence that an initiating event cannot occur.

Robustness concepts also need to be applied to secondary structures. Secondary structures such as balconies, handrails and staircases are often designed by subcontractors and the principal engineer may be less involved (either through contractual reasons or programme constraints), and therefore the impact of the failure of a secondary structure on the building may not be considered.

>  *CROSS Safety Alert: Tension systems and post-drilled fixings*[23]
> Lining failures occurred in the Boston Big Dig tunnel in 2006, in the Balcombe rail tunnel in the UK in 2011 and in Japan's Sasago tunnel in 2012; all related to the use of unsuitable fixings.

2.11 A robust building is tied

A robust building is built of elements that are effectively 'tied' together, and do not rely on bearing or frictional connections.

A house of cards built from cardboard playing cards stays up because of the friction between the fine edge of a card and the flat face of the supporting card (Figure 2.12). This is obviously not a robust structure.

If the cards were taped at their intersections, the playing card structure would be far more likely to stay standing up. The tape is acting like a 'tie' in a building.

Figure 2.12: House of cards

The provision of building ties is the prescriptive approach described in BS EN 1991-1-7[4].

The Ronan Point collapse[24] (Figure 2.13) is the best-known building collapse in the UK, and it is the response to this collapse that led the UK to put together the building standards for robustness and disproportionate collapse and a revision of the building regulations from 1970.

Figure 2.13: Ronan Point collapse

Mrs Ivy Hodge lived in a corner flat on the 18th floor of Ronan Point in Newham, East London. In May 1968 she lit a match to light her gas stove for a cup of tea. The match sparked a gas explosion that blew out the loadbearing external walls, which had been supporting the four flats above.

The corner of these four flats crashed onto the flats below and the entire corner of the building collapsed; four people died and 17 were injured.

Ivy Hodge was not injured. She even took her gas cooker to her next home. This is significant. The fact that Ivy Hodge survived implied that the explosion was not particularly large. In fact, a dent in her kettle reportedly led to the conclusion that the explosive force was approximately $34kN/m^2$, a very unscientific conclusion but one that is still relied upon today.

The building collapse was therefore disproportionate to the accidental event; in this case a relatively minor gas explosion.

Ronan Point was built of precast wall panels and floors. It was discovered that the structural components were not adequately tied together and, once an external wall was knocked out, there was nothing to prevent collapse of the structure above (Figure 2.14). The weight of the falling structure knocked out more floors below.

Figure 2.14: Ronan Point floor-to-wall interface

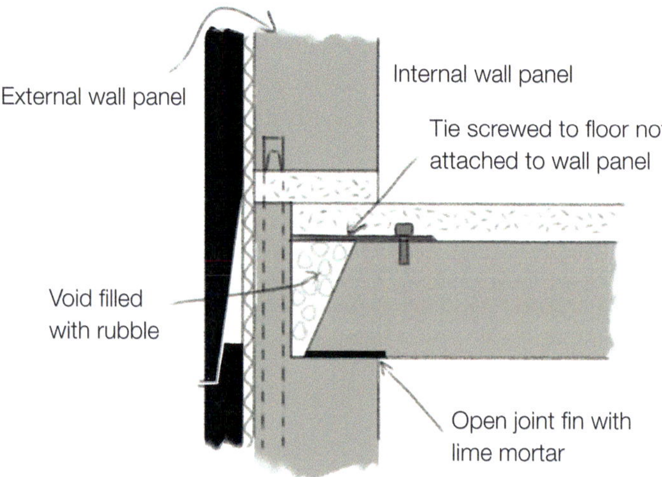

Although this chapter has discussed many concepts of robustness, Approved Document A[25] of the Building Regulations[26] puts it very simply (Table 2.1).

Table 2.1: Approved Document A, disproportionate collapse definition

Requirement
Disproportionate collapse **A3.** The building shall be constructed so that in the event of an accident the building will not suffer collapse to an extent disproportionate to the cause.

Detail on how A3 is achieved is set out in Section 5 of Approved Document A, BS EN 1991-1-7 and the associated material codes. Key to applying BS EN 1991-1-7 is for the engineer to understand the risks associated with the building. Requirement A3 is Statute within Schedule 1 of the Building Regulations.

 *CROSS Safety Report: Collapse of large panel system (LPS) buildings during demolition[27]*
In 2009, numerous progressive and unexpected collapses occurred during the demolition of large panel system (LPS) tower blocks due to lack of ties between floors and cross-walls.

2.12 Progressive collapse

The Ronan Point failure was a classic example of progressive collapse; that is the failure of one member which set off a chain reaction of other collapses, such that the totality of damage was disproportionate to the initiating event.

No engineer can prevent total collapse if the event is big enough, but a robust structure should ensure that the extent of damage is not disproportionate to the initiating event.

A more recent example of this is the Morandi Bridge collapse in Genoa, Italy in 2018[28] (Figure 2.15), where the failure of one stay due to corrosion caused collapse of a whole section of the structure.

Figure 2.15: Morandi Bridge collapse

Another, recent example is that of Champlain Towers South in Miami. Champlain Towers was a 12-storey building which collapsed suddenly in June 2021 with 98 fatalities. It was of reinforced concrete construction, approximately 40 years old, and was located adjacent to the ocean.

The most complex forensic investigation ever undertaken in the US is being carried out by NIST (National Institute of Standards and Technology) and at the time of writing is still ongoing.

Clearly a lack of robustness was a factor in this extreme case of progressive collapse, but the causes are not yet known. The NIST results and conclusions will undoubtedly have a profound influence on how ageing concrete buildings are viewed in the US and around the world.

2.13 Consequence class

Key to designing for robustness is to determine the consequence class (CC) of a building. There are four consequence classes; CC1, CC2a, CC2b and CC3 and each represents an increasing level of impact on occupants if there were to be a full or partial collapse, and also an increase in the extent of collapse if robustness is not considered. CC3 is outside the scope of this *Guidance*. Consequence classes are further discussed in Sections 3.3, 4.3, 5.1 and Chapter 6.

3 Risk

3.1 Introduction

Engineers subconsciously consider risk all the time but may not realise that the risk assessments made could have a huge impact on society. Risks are also assessed within the culture and history of the society in which they are working. One of the most common causes of damage to a building is vehicular impact. This risk seems to have been accepted by everyone; vehicles have not been banned and crash barriers have not been erected alongside every pavement. However, if the roof of a school were to collapse and cause multiple casualties, people would, quite rightly, be horrified.

People also tend to be more scared by large-scale building collapse rather than secondary elements of a building failing. In Nepal, after the earthquake in 2015, the media concentrated on the wholescale collapse of buildings, but a UK engineering team, assessing damaged buildings, was concerned about the rooftop water tanks falling from the roofs onto people in the streets. Individuals are more concerned about risks that they do not understand than risks that they do. With the latter, a person feels that they can make an informed decision to keep themselves safe.

An element of how risk is perceived is based on how society has grown up and how cities have developed. If a new building was designed which vehicles could get very close to, the decision would be made either to keep them away, design the building to cope with impact load or protect the building with impact barriers. However, if it were planned to refurbish the historic market building in Tetbury, for example (Figure 3.1), crash barriers would not be retrofitted to the columns. Different decisions are made based on culture and expectations.

Figure 3.1: Market building, Tetbury

Engineers designing buildings to the rules set out in the Eurocodes, are designing for a risk evaluation that is implicit in the codes. Therefore, they are not necessarily thinking about actual risks or the structural performance of the building. This may be acceptable for standard buildings and typical structural forms, but not for all buildings. Under the Building Safety Act[2], existing higher-risk buildings will require a safety case report, and this may be the trigger for structural engineers to consider risk in greater depth.

3.2 Risk and robustness

Achieving robustness requires engineers to understand risk. Possible initiating events need to be considered and each one assessed to determine the likelihood of occurrence. It is necessary to determine how a building might respond to this event and decide whether the consequence of the response of the building is acceptable. In December 2021 CROSS published a Structural Alert[29] which stated:

"Structural and civil and fire engineers work in a high risk environment; most of the time all is well, but when a problem occurs it can be very serious both in terms of life, commercial cost and reputation.

Managing safety risk is not an optional exercise. It is a statutory obligation, will be a contractual requirement, and is an obligation under Institution Codes of Practice. It requires a disciplined approach by the designer, on all projects (risk does not respect project size)."

Eurocode 1 makes a distinction between designing buildings for known risks (identifiable accidental actions) and designing buildings where risks cannot be identified and where the strategy is for limiting localised failure. Figure 3.2, derived from BS EN 1991-1-7[4] shows these different strategies.

Figure 3.2: Strategies for accidental design situations

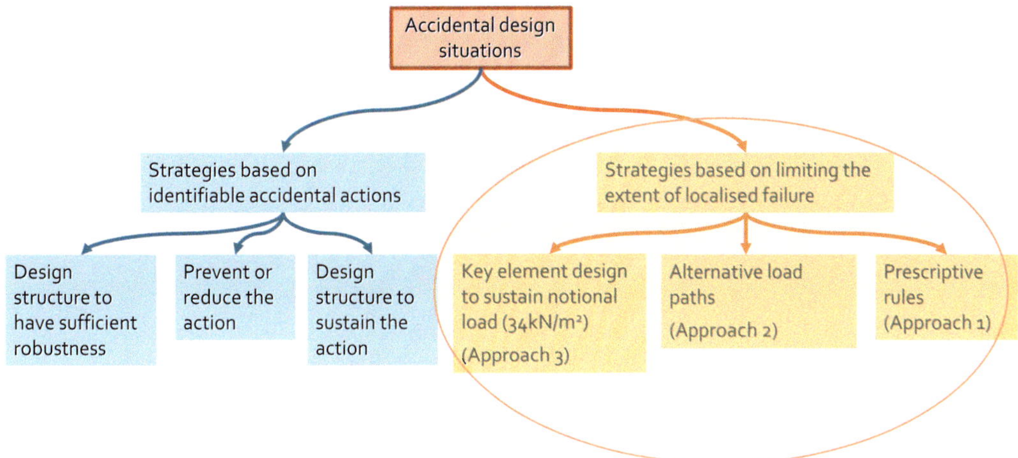

Examples of identifiable risks are:

- Car impact in an underground car park
- Lorry impact in a delivery bay
- A railway line very close to the building with possible train derailment
- An industrial process that could cause an explosion
- Electric car fire in an underground or multi-storey car park

An unidentifiable risk, or unqualified risk, might be a gas explosion in a residential building.

This might sound odd — the term 'gas explosion' suggests that the risk has been identified, but the point is that the provision of gas in buildings is considered part of everyday life and not considered extraordinary, even though many houses have been destroyed by gas explosions.

It also seems sensible to design for a baseline level risk; even if there is no gas in a building there is always a possibility of an explosion. Note, however, that after Ronan Point, gas was removed from many flats of the same type of construction. This baseline level risk is a pressure of $34kN/m^2$ and although it was derived from a gas explosion, designing for this gives a level of robustness for a whole range of potential unidentified risks. It is used as the accidental load in the design of key elements unless accidental forces are known to be higher.

There will need to be increased discussion about risks in higher-risk buildings (HRBs), which include many large panel system (LPS) buildings.

In England there are an estimated 12,500 HRBs which will be prioritised and assessed over the five-year period from 2023 for structural safety and spread of fire.

The presence of gas (there are still some HRBs with mains gas), is just one of the potential risks.

This *Guidance* is mostly about designing for unidentifiable risk, but at all times as an engineer, it is important also to consider identifiable risks.

Engineers frequently need to take a 'mix and match' approach, designing for both identifiable and unidentifiable risks.

A typical example of having to take such an approach is in the design of a residential apartment building constructed over a car park. The structure would be designed for the unidentifiable risks but would also be designed for vehicle impact loads on the columns within the parking area (Figure 3.3).

Figure 3.3: Design strategies for an apartment building over a car park

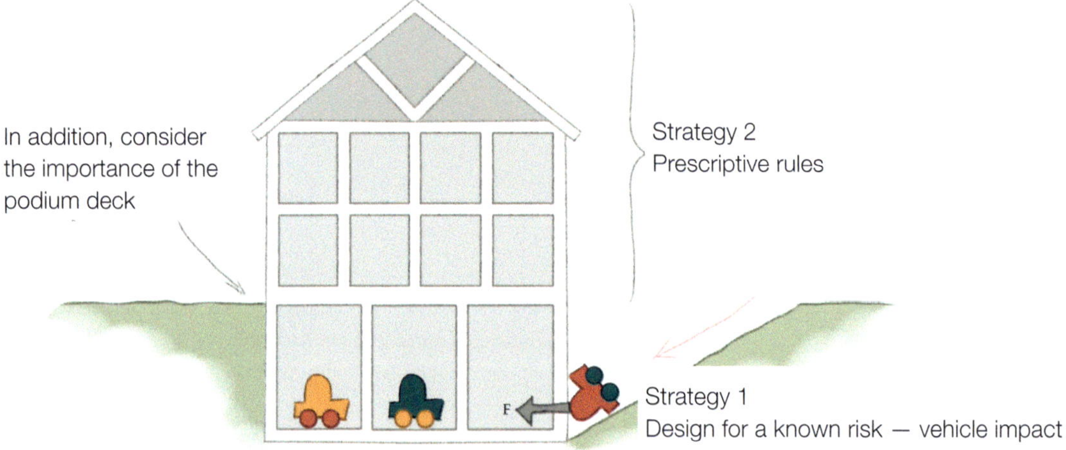

If the column layout is different in the car park storey from that in the upper storeys, the podium deck becomes a transfer structure and is key to the ability of the building to cope with an accident.

A less common example is a building next to a railway line, where the design engineer decided to move the building's lower floors away from the railway line in case of train derailment, and proceeded to design the building for unidentifiable risks (Figure 3.4).

The nature of a structure and its layout also has an impact on the risk that a building poses to human life. The more complex a building, the more likely it is that any accidental and unexpected loading will cause collapse. In the building shown in Fig. 3.4, although the lower floors were moved away from the railway line, the upper floors cantilevered out by creating a truss between second and third floors. This truss was absolutely critical to the structure above and each part of it was designed as a key element. Above the truss, the building was designed using prescriptive tying rules.

Figure 3.4: Design strategies for a building next to a railway line

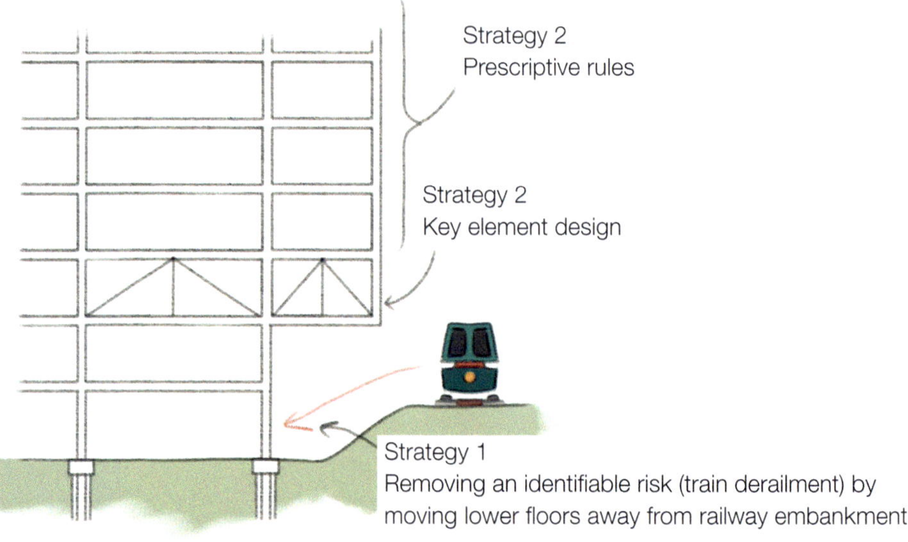

3.3 Risk and Approved Document A

In Approved Document A[25], risk with respect to disproportionate collapse is first mentioned in Section 5.1e which states:

"For Consequence Class 3 buildings — A systematic risk assessment of the building should be undertaken taking into account all the normal hazards that may reasonably be foreseen, together with any abnormal hazards."

Approved Document A is advising that robustness of a CC3 building should be determined by carrying out a risk assessment. In 2013 the IStructE published the *Manual for the systematic risk assessment of high-risk structures against disproportionate collapse*[19] which describes and discusses the risk assessment process. This is shown in the flowchart in Figure 3.5.

Figure 3.5: Risk assessment process for CC3 buildings

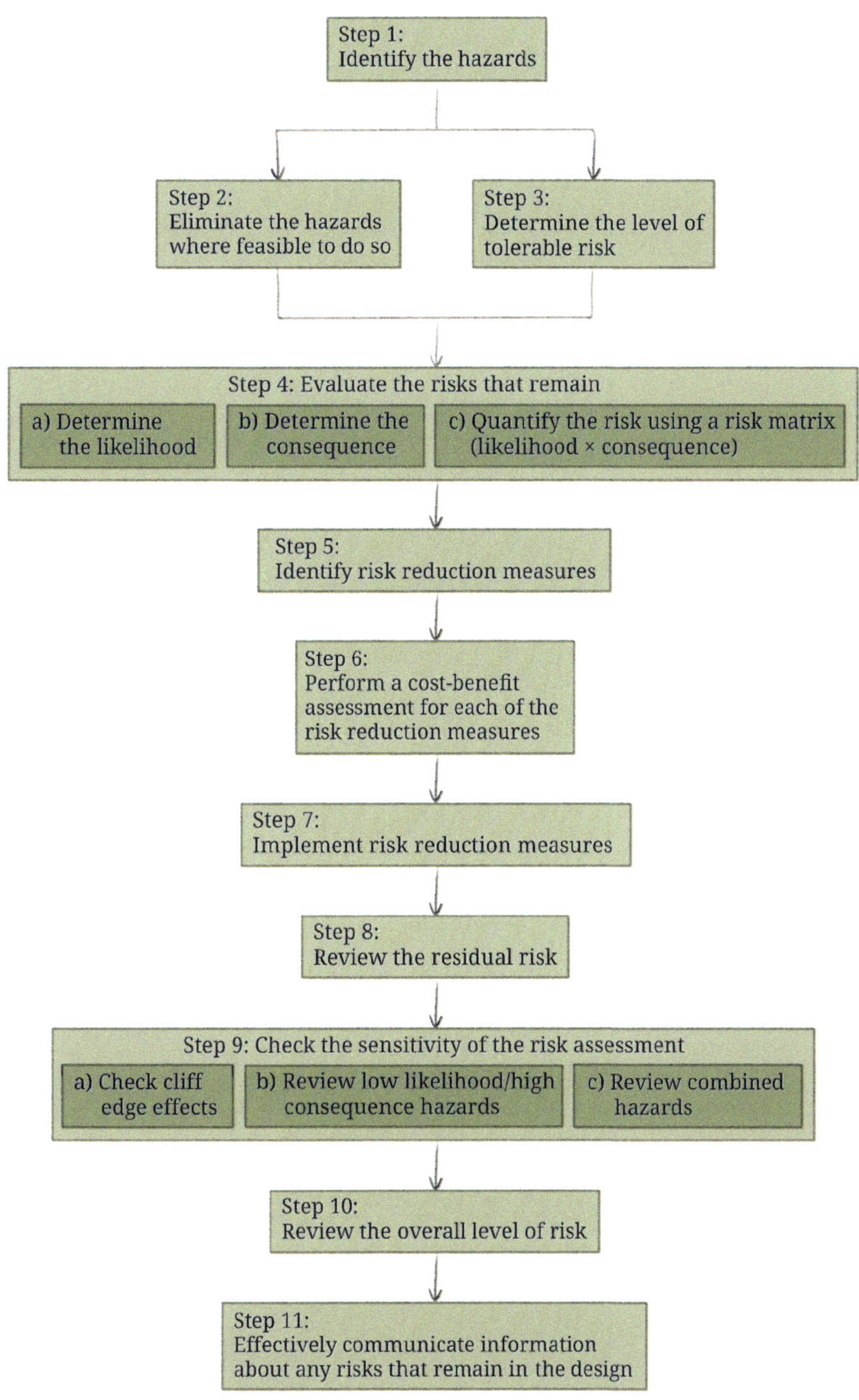

The second reference to risk is more oblique and is in Section 5.4 of Approved Document A:

"As an alternative to Table 11, for any building which does not fall into the classes listed under Table 11, or for which the consequences of collapse may warrant particular examination of the risks involved, performance may be demonstrated using the recommendations given in the following Reports and Publications…"

The reports and publications referred to include a DTLR document[30] based on an Allott and Lomax proposal on how to assess risk. This document identifies a number of risks:

C: Human occupancy
S: Societal parameter
E: Environment parameter
D: Structural parameter

Allott and Lomax were also involved in the design of the consequence classes which are based on these parameters. The societal parameter is very interesting because it is intended to reflect the perception of society on the impact of an accident or action occurring in different buildings.

The report says:

"Clearly events which affect large gatherings of people in circumstances that are not unusual are likely to be perceived far more seriously, i.e., with a high consequence, than those which affect a small minority in rare or unusual circumstances."

Engineers will have noticed that school buildings are classified as CC2b at just two storeys in height. This reflects societal perception; the deaths of children shock people more than the deaths of adults. Children are also more vulnerable in that they have less ability than an adult to deal with an incident or accident. Unfortunately, schools have not had a good record of robustness over the last few decades.

3.4 Designing for risk

Risks to a building include potential accidental loads, unknown accidental loading and also its structural layout. All of these aspects need to be considered in the design of a robust building. For smaller structures it would be wise to undertake an assessment of risks to a building even if very informally. However, the Building Safety Act makes it a legal requirement to carry out regular risk assessments for HRBs and this should be done in a formal way, by experienced engineers within a culture of peer review and checking.

The Allott and Lomax categories may be a useful framework for looking at risks. A risk-based approach to designing a robust building may also be the most appropriate way of designing alterations to existing buildings.

CROSS Safety Alert: Hazard identification for structural design[31]
At the outset of the design, it is recommended that the structural engineer should document the proposed structural design philosophy and the approach to design against disproportionate collapse and agree this with the client.

Risk in structural engineering[32]

4 Regulations, codes of practice, resources and supporting documents

4.1 Introduction

The key Eurocode for the design of a robust building is BS EN 1991-1-7[4], Eurocode 1 — Actions on structures — Parts 1–7: General actions — Accidental actions. When designing in any material, it is important to reference this code.

The code describes principles and application rules for the assessment of accidental actions on both buildings and bridges. It includes impact forces from vehicles, rail traffic, ships and helicopters. It also includes actions due to internal explosions and actions due to local failure from an unspecified cause. There is an accompanying National Annex for the UK[33].

It does not specifically deal with accidental actions caused by external explosions, warfare and terrorist activities, or the residual stability of buildings damaged by seismic action or fire.

Robustness is defined in Section 1.5.14 of the code as:

"The ability of a structure to withstand events like fire, explosions, impact or the consequences of human error, without being damaged to an extent disproportionate to the original cause."

Section 2 of the code sets out succinctly the various subclauses within BS EN 1990[34] which address accidental actions. Section 3 is effectively the starting point for designing a robust building to avoid disproportionate collapse.

4.2 Design strategies and approaches to prevent disproportionate collapse

Section 3 of the code sets out the two different strategies for designing a robust structure. The first strategy is when the possible accidental actions and risks are known and identifiable, and the second strategy is when they are unknown and unidentifiable (Figure 3.1 in BS EN 1991-1-7).

Where accidental actions are unknown, the approaches are based on limiting the extent of localised failure and one or more of three approaches can be taken:

- Notional element removal — provide enhanced redundancy by examining the behaviour of a building if structural elements are removed one at a time
- Key element design — design structural elements as key elements to sustain a notional accidental action. This notional accidental action is $34kN/m^2$ applied as horizontal load
- Prescriptive rules —follow prescriptive rules, and provide vertical and horizontal ties, the capacity of which depends on building class and material

The two strategies:
- Identifiable risks
- Unidentifiable risks

The three approaches:
- Notional element removal
- Key element design
- Prescriptive rules

It is very common to design a building using a mix of approaches and even both strategies. It is also pointed out in Section 3 of the code that there are some buildings where full collapse of a building could be allowed, but only should there be no risk to human life.

Within the code, the first strategy is explained further in Section 3.2 and reference is made to risk assessment, stating that the accidental actions that should be considered depend on:

- Prevention or reduction of accidental actions
- Probability and consequence of the action
- Public perception
- Level of identifiable risk

It is interesting to see that public perception is mentioned; this suggests that structural engineers subconsciously make different design decisions depending on who they are designing for.

Removing a risk is always a good starting point for building design. If a building is close to a hazard, such as a road, it should be investigated whether the building can be moved away from the road, or the road moved away from the building or a barrier introduced.

The second strategy is set out in Section 3.3 of the code. The key element approach is discussed in 3.3 (2) a), the removal of structural elements approach is set out in 3.3 (2) b) and the prescriptive approach in 3.3 (2) c). Detail of how to calculate tie forces for both framed buildings and loadbearing wall construction is set out in Annex A (A.5 and A.6). However, some of the materials use different tie forces that are set out in the specific material codes and 'non-contradictory complementary information' guidance notes.

4.3 Consequence classes

It is also Section 3 of the code that introduces the concept of consequences classes for buildings. The definition of each consequence class is covered in 3.4 (1) and the way accidental design situations should be considered is covered in 3.4 (2) (Table 4.1). Full details on consequence classes, and what type of buildings fall into each class, is set out in Table A.1 of Annex A.

Table 4.1: Definitions of consequence classes

Consequence class	Definition (from 3.4 (1))	Considerations (from 3.4 (2))
CC1	Low consequences of failure	No specific consideration is necessary for accidental actions, except that the robustness and stability rules given in EN 1990 to EN 1999 as applicable are met
CC2	Medium consequence of failure	Depending on the specific circumstances of the structure, a simplified analysis by static equivalent action models may be adopted or prescriptive design/detailing rules may be applied
CC3	High consequences of failure	An examination of the specific case should be carried out to determine the level of reliability and the depth of structural analysis required. It may be necessary for a risk analysis to be carried out and the use of refined methods, such as dynamic analyses, non-linear models and interaction between the load and the structure

4.4 Vehicle impact

Section 4 of BS EN 1991-1-7 deals specifically with impact, describing events such as impact from road vehicles, forklift trucks, trains, ships and the landing of helicopters on roofs.

Actions for impact should be considered for buildings used for car parking, buildings in which vehicles or forklift trucks are permitted, and buildings that are located adjacent to either road or railway traffic. Buildings where the roof contains a designated landing pad should be designed for actions from the impact of helicopters.

Section 4.2 of the code states that actions due to impact should be determined by a dynamic analysis or represented by an equivalent static load. Annex C includes the formulae for dynamic design for impact loading and covers both vehicles and ships. Annex C also delineates between hard impact and soft impact. Hard impact is when the building structure is rigid and the colliding object absorbs the energy. Soft impact is the opposite and is when the building can deform and absorb the energy but the colliding object is rigid.

It is unusual for design engineers to look at dynamic loading when it comes to building design, and therefore Section 4 of the code will be more commonly used because it includes the equivalent static impact loads for vehicles, ships and helicopters.

4.5 Explosions

Section 5 of the code covers internal explosions but makes the point that blasts due to explosives are outside the scope of the code. Instead, it covers explosions in buildings such as chemical facilities or sewage constructions where:

- Gas is burned or regulated
- Explosive gases or liquids are stored or transported

The section makes the statement that there does not need to be specific consideration of the effects of an explosion, provided that the rules for connections and interactions between components provided in BS EN 1992 to BS EN 1999 are complied with for a building of Consequence Class 1. However, it is very unlikely that a CC1 building will house gases or explosive liquids.

Section 5 of the code is mostly relevant to CC2 and CC3 buildings, with the further requirement that CC3 buildings should be designed using a dynamic analysis.

Annex D of the code provides formulae for calculating actions due to:

- Natural gas explosions
- Dust explosions
- Explosions in road and rail tunnels
- Explosions in ducts

However, it is important to point out that PD 6688-1-7[35], which gives complimentary and non-contradictory information on BS EN 1991-1-7 and its National Annex recommends that Annex D is not used in the UK.

4.6 Annexes

There are four annexes in BS EN 1991-1-7. A, C and D have already been mentioned but Annex B has not been discussed.

Annex B provides information on risk assessment and could be a source for building design when designing a CC3 building or buildings with the possibility of unusual loads.

This *Guidance* does not cover CC3 buildings, but, as has been stated in Chapter 2, the principles of risk assessment should be applied to all buildings, even if in a very simple way.

4.7 The National Annex to BS EN 1991-1-7

The National Annex gives UK decisions for the Nationally Determined Parameters described in many of the clauses in BS EN 1991-1-7. Therefore, BS EN 1991-1-7 should always be read in conjunction with the National Annex.

4.8 Use of BS EN 1991-1-7 in conjunction with other codes of practice

BS EN 1991-1-7 is the key Eurocode for designing to prevent disproportionate collapse. However, it is important to consider the Eurocodes for each material and all National Annexes.

The steel code, Eurocode 3[36] contains no specific requirements for robustness, so BS EN 1991-1-7 is the only source. BS EN 1991-1-7 is also applicable to light steel frames, setting out additional guidance for lightweight construction in the UK National Annex to BS EN 1991-1-7, Clause NA.3.1.

The concrete code, Eurocode 2[37], contains material-specific formulae for checking horizontal and vertical ties, with the values being set out in its National Annex[38]. Some of these values are in line with those in the withdrawn British Standard, BS 8110[39], but the design engineer should not rely on memory and should always refer to the Eurocode.

In the masonry code, Eurocode 6[40], there is nothing about disproportionate collapse and the design engineer should refer to PD 6697[41].

In the timber code, Eurocode 5[42], tie forces are not mentioned, so the engineer should refer to PD 6693-1[43], the complementary and non-contradictory guidance for the UK. A timber-framed building would be considered lightweight construction (as is light steel frame), and PD 6693-1 allows a lower set of tie forces to be used.

Approved Document A of the Building Regulations[25] contains a section on robustness and disproportionate collapse which effectively overlaps with that of BS EN 1991-1-7.

4.9 BS EN 1990

When designing for robustness, the engineer needs to use appropriate load combinations and load factors. Reference should be made to Expression 6.11b in the code[34] and Tables NA.A1.1 and NA.A1.3 in the National Annex to BS EN 1990[44]. This is explored further in Section 5.8 of this *Guidance*.

4.10 *Review of international research on structural robustness and disproportionate collapse*

This document[45] was published by the Department for Communities and Local Government (DCLG) and the Centre for the Protection of National Infrastructure, and authored by Arup.

Published in October 2011, it is still probably the most extensive investigation into the current approaches for designing a robust building and also the history of the legislation. It also considers approaches taken by other countries.

It is essential reading for a design engineer involved with a more challenging building, and is recommended reading for anyone who wishes to understand the history and sources of robustness rules and recommendations.

The review was published one year after the first edition of this *Guidance* and seven years after the 2004 changes to Approved Document A. Therefore, it is currently the most up-to-date commentary on robustness and disproportionate collapse.

4.11 *Manual for the systematic risk assessment of high-risk structures against disproportionate collapse*

This publication[19] is a must-have document for design engineers, especially when designing a CC3 building. In Chapter 2 it makes the important point that a systematic risk assessment is not a substitute for a robust structural form; something that the design engineer must remember.

The book also covers the concepts of 'So far as reasonably practicable' (SFARP) and 'As low as reasonably practicable' (ALARP). Understanding these concepts is increasingly important now that the Building Safety Act[2] requires all existing higher-risk buildings to have a fire and structural safety risk assessment.

The last chapter is devoted to works to existing buildings, and therefore it is strongly recommended that any engineer who is involved with the design of alterations to existing buildings has access to this book.

4.12 CROSS

Collaborative Reporting for Safer Structures (CROSS)[46] is a confidential reporting system that captures — and shares — lessons learnt from fire and structural safety issues.

CROSS currently operates in the UK (CROSS-UK), Australasia (CROSS-AUS) and the US (CROSS-US).

CROSS produces CROSS Safety Reports, Safety Alerts, Articles, Reviews and Topic Papers and enables engineers to investigate and understand building failures. Understanding how a building may fail or collapse helps inform design. Some of the CROSS Reports and Alerts are signposted in this *Guidance* for further reading.

4.13 The Building Safety Act

The regulatory framework has changed in the wake of the Grenfell Tower tragedy and the Building Safety Act sets out the changes to the construction industry, including the Health and Safety Executive's (HSE) involvement in the design and construction and checking of higher-risk/'in scope' buildings (HRBs) and the creation of the Building Safety Regulator.

This Act introduces reforms to give residents and homeowners more rights, powers and protection, and to ensure that homes across the country are safer.

The expansion of CROSS to include fire safety as a result of the Grenfell Tower fire will also propel CROSS greater into the consciousness of the construction industry, and therefore the Building Safety Act is considered to be exciting legislation.

A series of gateways have also been introduced for HRBs that need to be passed through before the design and construction team can move on to the next stage of the process. Demands on designers, contractors and building owners will be greater and it is an opportunity to clarify some of the grey areas of building design, such as hybrid construction and uncertainty of design responsibilities.

A 'golden thread of information' will need to be created and maintained, and this will place responsibilities on structural engineers. The golden thread is both:

- Information about a building that allows understanding of a building and the ability to keep it safe
- An information management process to ensure the information is accurate, easily understandable, can be accessed by those who need it and is up-to-date

One key point about the changes is that a risk assessment will need to be carried out on new and already-occupied HRBs. A building safety risk is defined as the spread of fire, or structural failure that could cause significant harm to those in or around the building. Structural engineers will need to have a greater understanding of fire in buildings and the immediate and longer term effects of a fire on a structure. The HSE states that the building safety risk assessment should consider all potential scenarios for the building.

There are powers within the Building Safety Act to make regulations regarding competence requirements of those working within the built environment. These regulations impose a requirement on anyone carrying out any design work to be competent in their roles.

Guidance that has been published in recent years includes the competencies shown in Table 4.2 for the design of HRBs, and design engineers should anticipate something similar in the coming years when assessing existing buildings or designing new.

Table 4.2: Provisional competencies for structural engineers

Scope	Competence
Structural safety Compliance based ('deemed to satisfy') and first principles ('engineered') approaches to both fire engineering and design against disproportionate and progressive collapse	The ability to identify correctly areas of sensitivity in the design of the structure of an HRB, and to identify areas of sensitivity which might result in fire-related or structural safety risks
	Understanding of the change in behaviour of the structure if a structural or fire-related action were to be larger than anticipated, or the strength or stiffness were to be lower than anticipated following an event
	Understanding of the difference between performance-based and compliance-based principles in demonstrating against structural collapse and fire-related hazards
	The ability to select an appropriate design solution that addresses the identified major accident, fire and structural hazards, applying ALARP principles
	The ability to confirm independently the overall adequacy of the structural design for a scheme through order of magnitude checks, by way of peer reviews and detailed design checks
Structural design, disproportionate collapse and second-order effects	The ability to develop a structural design in which the identified major accident hazards relating to fire and structural safety are managed to ALARP principles
	The ability to design a structure which will behave predictably under fire and structural hazards, and remain stable after the event
	Understanding of the non-linear second-order effects and their impact on the structural safety of the building

Under these regulations, individual engineers will need to have the skills, knowledge, experience and behaviours necessary for the role being performed. The Institution of Structural Engineers and Institution of Civil Engineers both have codes of conduct that expect engineers to work within their competence, undertake continuous professional development and have full regard for public interest, particularly with respect to safety.

4.14 The Building Act 1984 and Building Regulations

Structural engineers work within the Statute of the Building Act[47]. The Building Act 1984 provides the Secretary of State the power to create Building Regulations[26]. The Statute of the Building Act is explored in Chapter 13 and the design engineer should be familiar with the Act, the Building Regulations and also Approved Document A which provides guidance on how engineers can design a building which complies with the Statute.

5 Designing for robustness

5.1 Introduction

This chapter describes the approaches required to achieve a robust structure. The principles discussed are applicable to all materials, although the manner of application is material-dependent and specific advice is given in the relevant material chapters.

The importance of considering robustness and safety from the earliest stages of a project is highlighted, followed by the need for clear direction to ensure the philosophy developed is implemented within the final design, detailing and construction.

For most structures, the standard means of achieving robustness is by compliance with rules in Approved Document A[25] and BS EN 1991-1-7[4]. The requirements are subtly different, so it is important to read both documents.

From these documents, the design criteria for the various classes of structure are as shown in Table 5.1.

Table 5.1: Design criteria for consequence classes

Consequence class	Requirements	Notes
CC1	Provided a building has been designed and constructed in accordance with the rules given in EN 1990 to EN 1999 for satisfying stability in normal use, no further specific consideration is necessary regarding accidental actions from unidentified causes	Once horizontal and vertical loading have been considered with respect to ultimate and serviceability design, no further design is required for robustness
CC2a	Provide effective horizontal ties or effective anchorage of suspended floors and roofs to walls, as described in the material codes	BS EN 1991-1-7 states that for this class of structure, horizontal ties should be used for framed structures, and anchorage of suspended floors and roofs can be adopted for loadbearing wall construction in conjunction with a cellular layout (A.5.1 and A.5.2 of the National Annex[33]) In some cases, it may also be appropriate to adopt horizontal tying for loadbearing wall construction
CC2b	The provision of horizontal ties, as defined in A.5.1 for framed buildings and A.5.2 for loadbearing wall construction, together with vertical ties, as defined in A.6 in all supporting columns and walls Alternatively the building shall be checked to ensure that upon the notional removal of each supporting column and section of wall, and each beam supporting a column, one at a time in each storey of the building, the building remains stable and any local damage does not exceed a certain limit Where the notional removal of such columns or wall sections would result in an extent of damage in excess of the agreed limit, such elements should be designed as key elements	Within BS EN 1991-1-7 and Approved Document A, horizontal ties are not specified as a requirement when the notional element removal approach is taken, or structure is designed as key elements In the DCLG report[45] there is the suggestion that horizontal ties were always intended to be included in a CC2b building, even if the alternative approaches were taken, and that the omission of this requirement may have been an error introduced in 2014
CC3	Outside the scope of this *Guidance*	

With regard to the DCLG report, it is a widely-held view that provision of horizontal ties is always required in CC2b buildings unless there is good evidence to the contrary. Structurally it is very difficult to understand how it is possible to achieve a successful result (limited area of collapse) from notional removal if there is no tying in the structure.

In some industries, however, it has not been the practice to provide any additional physical ties if the structure is proved to be robust enough via the notional element removal approach.

An example of this would be in timber engineering, where full-scale tests on timber-framed structures have proved that CC2b structures are adequate, provided certain detailing rules for floor anchorages are adopted (Chapter 12 and Figure 12.1).

Another example would be that of cellular, traditionally-built masonry buildings with regular and repetitive room layouts which will be robust structures, even if horizontal ties are not provided. It is worth noting that BS EN 1991-1-7 states that a cellular form of construction should be adopted for loadbearing wall construction.

The design engineer will need to use judgement; not all timber structures and masonry structures will be robust based on their layout alone, and notional element removal may result in greater collapse than allowable, and therefore the omission of horizontal ties would be dangerous.

While the intentions of the categories are clear enough, buildings come in a variety of forms, and interpretation is required, especially in structures of mixed form (hybrid structures) or where alterations are being made.

Figure 5.1 provides a flowchart to guide the user through the various robustness options for the building consequence classes. Although CC3 buildings are not considered in this *Guidance*, there is a widely-accepted view that they should meet the requirements for CC2b as a minimum standard, in which case the explanation of design to CC2b will be relevant.

Although this *Guidance* is predominantly concerned with achieving structural robustness within individual buildings, opportunities to improve overall robustness by protecting the structure from hazardous events should always be considered when these events are reasonably foreseeable — for example, the protection of columns from vehicle impact, ensuring gas supplies are separated or protected, or removing electric car charging points from inside underground car parks.

Robustness of a structure itself can be considered at two levels: the overall structural concept followed by the detailed provisions.

5.2 Structural concept

The structural form of a building will significantly affect its robustness. Traditional cellular forms with many loadbearing walls assure a sensible level of robustness because loss of any one wall will generally not lead to the collapse of a large proportion of the structure. In contrast, having a large span supported on an easily-dislodged and/or vulnerable single column would not be a robust structure.

At the concept stage, the layout of the building and its basic structural action will need to be developed. For any structure, there should be an assessment of the hazards and provision of clear paths for horizontal and vertical loads back to the foundations. Additionally, a robust structural concept will be one which avoids situations where damage to small areas, or failure of any single element progresses to widespread collapse.

This is the ideal, but there are occasions when reliance does have to be placed on single elements. Once this is recognised, robustness can be improved by making such elements and their connections substantial.

Experience has shown certain arrangements to be potentially vulnerable, for example:

- Transfer beams — these are single beams which support a number of columns or hangers
- Elements considered individually to be minor that are required to ensure the stability of more significant elements
- Podium decks which are supported on a wider column layout than the structure over
- Cantilevers
- Long-span, simply-supported beams

Figure 5.1: Disproportionate collapse flowchart

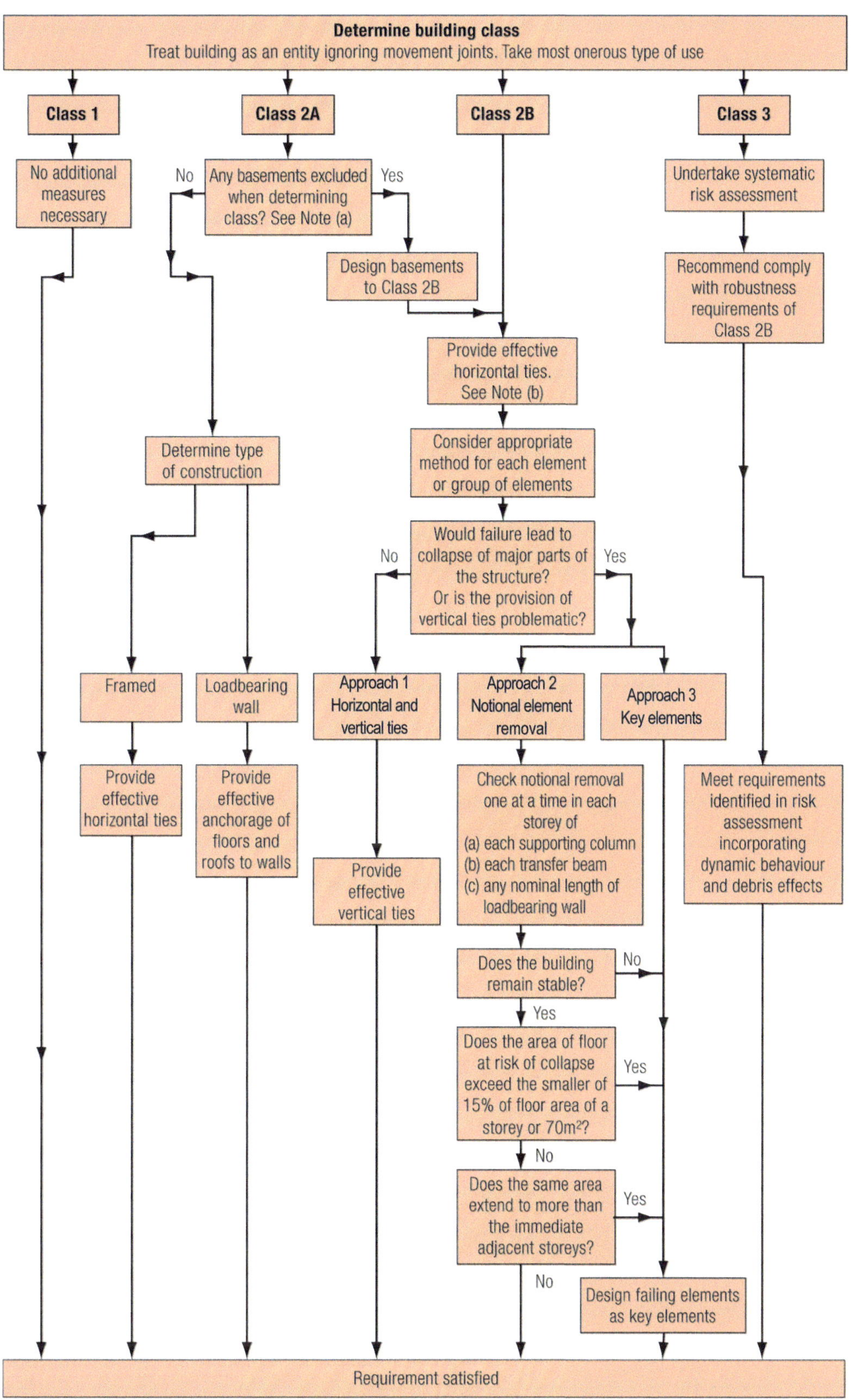

Notes

(a) Rules on the exclusion of basements vary in the building regulations of the various regions. When designing to EN 1991-1-7 basements are included in the storey count. However in England and Wales and Scotland and Northern Ireland, if they satisfy the Class 2B requirement, they may be excluded from the count.

(b) In Class 2B, for horizontal ties, rather than using physical ties in some cases alternative methods may be demonstrated by test.

The last two forms have no redundancy but that does not necessarily imply unacceptable vulnerability.

Some judgement is required on whether the element considered is significant, e.g., failure of a transfer beam supporting a single storey at high level is not as critical as one supporting several building storeys.

Functional restrictions on structural form may limit the scope of a designer to avoid arrangements that are potentially non-robust. However, early identification will allow the development of a strategy to provide the required compensatory robustness within the element itself. Once highlighted, a significant transfer beam could be designed to withstand a certain event (Section 5.7), or redundancy could be built into bracing systems such that the loss of a set number of braces does not permit restrained elements to buckle, or the structure as a whole to be placed at risk.

During project planning, it should be acknowledged that element detailed design may be carried out by someone other than the lead designer. This does not remove the need for the lead designer to take responsibility for design compatibility of the various elements, including responsibility for overall robustness. As a minimum, the overall robustness concept for the building, and therefore the effect that this will have on the design of individual elements, should be documented for later phases of the design development (which may be carried out by others). The Building Safety Act[2] and the golden thread of information requires that all design documentation should follow the project through all its life stages.

It is very common for a value engineering exercise to be carried out on projects, and one material, product or system substituted for another, with the substitution contractor-designed. A frequent example of this is *in situ* concrete columns being substituted for precast columns once the main contractor is appointed. This could affect robustness if the robustness strategy is not communicated to all of the design team.

Therefore on completion of the design, the structure should be checked to confirm that the initial assumptions are satisfied. This may include confirming that loads and load paths remain as anticipated, and that the overall robustness concept has been achieved.

The innate robustness of a cellular building applies to traditional masonry but may not apply to large panel system (LPS) buildings, particularly those from the 1960s and 1970s. Engineers should take special care when working on historic LPS buildings.

CROSS Safety Report: Concerns over risky new buildings?[48]
This report is about concern over cantilevers in buildings. They are often safety-critical because of their lack of redundancy and should be adequately tied back and supported. There are also concerns about future alterations to a building if the structural engineering team for the alterations is not aware of cantilevers within a building.

5.3 Horizontal loads, notional inclination and eccentricity

All structures should be capable of carrying horizontal loads applied in any direction on a horizontal plane, and there should be a defined load path for these loads back to the foundations. Wind action will generally suffice as a load to assess stability. However in some structures, wind loading is absent or not obvious, and in others it may be insufficient.

Historically in the UK, the original concept of notional horizontal load was to ensure that low-rise buildings which were subject to very low wind loads had capacity to resist unexpected accidental loading.

In the Eurocodes, the horizontal load is developed from a notional out-of-verticality/sway imperfection (inclination) and is a function of vertical load and, therefore, present whenever vertical load is present. In steel and concrete frame design, these equivalent horizontal loads are added to the wind loads, with the appropriate load combination.

The load combinations considered do include a percentage of wind loading in the accidental load case combination. At first consideration this seems improbable since accidental loading is unlikely to coincide with high wind loading, but if the notional loading is related to out-of-verticality, its inclusion becomes more rational. The effect is to ensure structures are robust against the effects of construction tolerance deviations, subsequent settlements or accidental horizontal loads.

In masonry design, both eccentricity due to construction imperfection and eccentricity due to the position of the vertical load are considered when determining the strength capacity of a wall. When undertaking alterations to existing buildings, it is common to hang new floors from a wall plate bolted to the side of an existing wall. However, the wall will have much less capacity to resist the weight of the new floor and the associated variable loads, than if the floor is built into the wall. These details, which may seem low in importance, could be key to the robustness of a building, and it is important that the engineer communicates these details to the contractor.

In timber structures, the bending stresses due to initial curvature, eccentricities and induced deflection are taken into account, and again, the detail of a connection, e.g., in a truss, could be key to the performance of that connection. A bolt that is positioned differently from that intended could seriously reduce the capacity of the joint. In platform-build, joists are sometimes hung from the inside of a stud wall or supported on the wall plate; these two alternatives induce different loads in the stud wall and it is important that the engineer understands how the timber frame will be built when designing.

5.4 Detailed provisions

Detailed provisions for robustness are given in material-specific codes of practice and described in Chapters 7–12 of this *Guidance* but there are typically three approaches:

- Approach 1: The indirect design method — provision of horizontal and vertical ties
- Approach 2: Alternative load path method — notional removal of elements
- Approach 3: Specific load resistance method — the provision of key elements

It is emphasised that these separate approaches are largely based on judgement — providing a level of robustness commensurate with routine risks, and achievable at affordable cost. It is not difficult to identify mathematical inconsistencies in them, but the sufficiency of the proposals has been proved over time. Potential accidental actions and even building forms change over time and their relevance to any particular design should be considered.

Approach 1 — for some buildings of CC2a and CC2b, resistance to progressive collapse is achieved by providing ties. The requirements for CC2b buildings are more onerous than for CC2a buildings and differ for each material. Adopting this method should provide buildings with sufficient robustness to survive a reasonable range of undefined accidental actions.

Approach 2 — the alternative path method presumes that through accidental loading a critical element is removed, and the structure is thereafter required to redistribute its gravity loads to the remaining structural elements via alternative load paths. There is no requirement in the UK to consider dynamic loads associated with the element removal. In practice, elements are notionally removed one-by-one and the residual structure (members and connections) tested for strength. Local collapse is not prohibited but its extent must not exceed prescribed limits set out in Approved Document A and BS EN 1991-1-7; that of 15% of the floor area or 100m^2, whichever is lower.

Approach 3 — certain elements are designed to sustain a notional upper bound to the abnormal loading (a value of 34kN/m^2 imposed pressure is used, derived from the Ronan Point blast); the presumption being that such members are strong enough to cope with a range of events.

The load factors for checking the ability of the building to resist the loading should be obtained from BS EN 1990[34]. The case studies in this *Guidance* give further clarification on load factors.

It should be noted that Approaches 2 and 3 are principally concerned with vertical structure or elements supporting vertical structure. When applying these approaches the designer must still ensure that the horizontal structure is robust in both directions. This is generally achieved by providing horizontal ties. There is confusion in BS EN 1991-1-7 where the implication is that if Approach 2 or 3 is taken, horizontal ties are not required. Refer to Table 5.1 in this *Guidance* for further explanation.

5.5 Approach 1: Tying

The provision of ties between floors and walls or columns, with a defined capacity, helps to constrain the elements from displacement during an extreme event. The provision of horizontal ties also makes alternative load carrying systems possible if a loadbearing element is lost, such as catenary and vierendeel action (Section 5.6.1).

5.5.1 Horizontal ties

The demand for horizontal ties differs between Consequence Class 2a and 2b and for different materials. In CC2a, it is sometimes possible to apply effective anchorage of suspended floors to walls, but the individual material codes should be checked.

In CC2b buildings, cross ties should be provided. The horizontal ties are split into three categories:

- Ties around the structure, called 'peripheral ties'
- Ties across the structure, called 'internal ties'
- Ties connecting horizontal structure to vertical loadbearing elements, which also need to be provided

Along a particular load path (which must be continuous) different structural elements , e.g., a series of beams, may be used as the ties, providing they have adequate interconnection. Rules for tie location and for their design forces are given in the material codes.

Ties can act to prevent the structure being dislodged, which is particularly important when supports are narrow, or at a perimeter where the ties must be capable of resisting any outward force on the supporting vertical element. The importance of such tying was demonstrated in the Ronan Point collapse where a gas explosion blew out a loadbearing wall, causing the slabs above to fall (Chapter 2). Lack of tying was also a factor in the Camden School collapse (Chapter 2).

Ties need to be continuous (i.e., lapped or connected) from edge-to-edge or around the structure, while at their ends, horizontal ties to edge columns and walls must be satisfactorily anchored back. All tie force paths should be geometrically straight; changes in direction to accommodate openings or similar discontinuities should be avoided wherever possible. Where such changes are unavoidable, the tendency of the tie to straighten under load should be considered and restraining elements provided. For buildings comprising separate structures, or incorporating joints creating structurally independent sections, the tie force requirements are applied to each independent section, with each treated as a separate unit.

Note that once tie forces are calculated, and ties and their connections provided, the structure that the tie is connected to does not need to be able to resist the tie force; the tie force does not need to be resolved. If it is calculated that a horizontal tie force needs to have capacity of 87kN, the structural element that the horizontal tie is connected to does not need to be checked on whether it can resist a horizontal force of 87kN. Although this seems illogical, it is simply part of the empirical approach.

The code-specified tie forces aim to ensure that slabs or beams can span across a removed support. There is no theoretical justification that this will actually happen if a column were to be removed. However, the implication of Approved Document A is that a notional tie provision represents a reasonable level of precaution. The consensus is that horizontal ties are an important safeguard which should always be incorporated and will protect most buildings against the majority hazards. Further consideration would be required for more demanding and unusual structural forms.

5.5.2 Vertical ties

Vertical ties have two roles. The first is to provide some form of minimum resistance to vertical elements being removed. The second is to enable load sharing between floors above a damaged vertical element. By linking a number of floors together, it is possible to provide a load path back to the intact structure above.

Rules on the calculation of the required capacity, and application of the vertical ties differ between materials and are set out in the material codes and BS EN 1991-1-7.

When required, vertical ties must be continuous from the lowest level to the highest level, and this includes anchorage into the footing or foundation. However, it is not necessary for the calculated tie force to be transferred from the ground floor column or wall to the foundation through its connection.

The rationale for tying into foundations is that this helps reduce the possibility of lowest column removal.

In structures where the lowest storey is a basement, if vertical ties through to the foundation are not possible, consideration of the basement structure as a key element normally provides a practicable solution (Section 5.7).

 *Historical review of prescriptive design rules for robustness after the collapse of Ronan Point*[49]

5.6 Approach 2: Notional element removal

An alternative to the tying method of providing robustness is to consider notional element removal. The term 'notional' is used deliberately as a means of emphasising an imaginary scenario.

In this approach, a defined element is removed (the element might be a column or nominal wall length) and the consequence examined. It is assumed that the act of removal itself does not induce any forces, static or dynamic. Equally, no benefit should be taken for enhanced material strength due to fast strain rate effects. After each removal, the building as a whole must remain stable and its members must not be destabilised by the removal, except for localised damage within the prescribed limits set out in Approved Document A and BS EN 1991-1-7. If such removal cannot be tolerated, the member must be designed as a key element.

What has to be hypothetically removed one at a time in each storey is 'each supporting column and each beam supporting one or more columns (usually called a transfer beam) or any nominal length of loadbearing wall'.

Loadbearing construction includes masonry walls and walls of close-centred timber or light steel studs. The nominal length removed is generally 2.25h where h is the storey height, but in the case of an external masonry, timber or steel stud wall, the length removed is that between 'vertical lateral supports' (usually wind posts or return walls).

For loadbearing walls at the corner, the notionally removed length of wall should be equal to the wall height in each direction, but not less than the distance between expansion or control joints. The design structural engineer should check each material code and Chapters 7–12 of this *Guidance* for specific requirements.

Critical cases for wall removal often occur where the plan geometry of the structure changes significantly, such as at abrupt decreases in bay size or at re-entrant corners, or at locations where adjacent columns/walls are lightly loaded, where the bays have different tributary sizes, and where members frame in at different orientations or elevations. Provided consideration is given to this approach in the design phase, it is usually possible to show that the unsupported structure can survive via some alternative structural system.

A weakness of Approved Document A, Section 5.1 is the case where support is provided by a number of closely-spaced columns. It would be irrational just to remove a single column, and the recommendation here is that all columns within a plan diameter 2.25h should be removed simultaneously.

An alternative load path analysis should be examined for each floor, one at a time. For example, if a corner column/wall section is specified as the removed element, one analysis is performed for removal of the ground-floor corner column/wall section, another is performed for the removal of the first-floor corner column, another is performed for the second-floor corner column, and so on.

If the designer can show that a similar structural response is expected for column/wall removal on multiple floors, the analysis for these floors can be omitted but the justification for not performing these analyses should be documented.

Following element removal, strategies for survival might be:

- Spanning the floor at right angles to its normal span using the available reinforcement (concrete floors) or boarding (timber floors)
- Making use of the available reinforcement in reinforced concrete floor to enable the element to span over the notionally removed column, if the floor is continuous over the column
- Using the ability of a wall to cantilever or span as a deep beam over the notional opening in the wall
- The provision of bridging elements such as those used in timber platform frame construction (rim boards)
- Catenary action of floors and beams
- Vertical and horizontal structural elements forming vierendeels over the missing structure

With analysis software, it is relatively simple to run several models of the same building but with different loadbearing elements missing, such as a corner column/wall, internal column and edge column, but with the usual caveat that the engineer should not use software to replace acquired knowledge and experience of building structures.

5.6.1 Catenary action

After element removal, a key survival strategy is to rely on the ability of the remaining connected members to span via catenary action, accepting that deflections may be high.

The principle of providing horizontal ties notionally allows for beam members to support loads by forming catenaries over damaged areas of a structure. The provision of horizontal ties, designed to Eurocode rules, has no complementary requirements relating to joint ductility or joint rotation capacity.

The strategy may rely on pure catenary action where the tension steel takes axial load only. More realistically, slabs and beams will retain some degraded moment capacity at their ends and centre, which will reduce the tie forces demanded. There is also a common case where the catenary receives some midspan support via tie forces from a surviving column above (Figure 5.2). Owing to the complexities in this, calculations cannot be accurate; they can only be crude approximations to suggest survival or failure.

Figure 5.2: Floor survival via catenary and hanger action

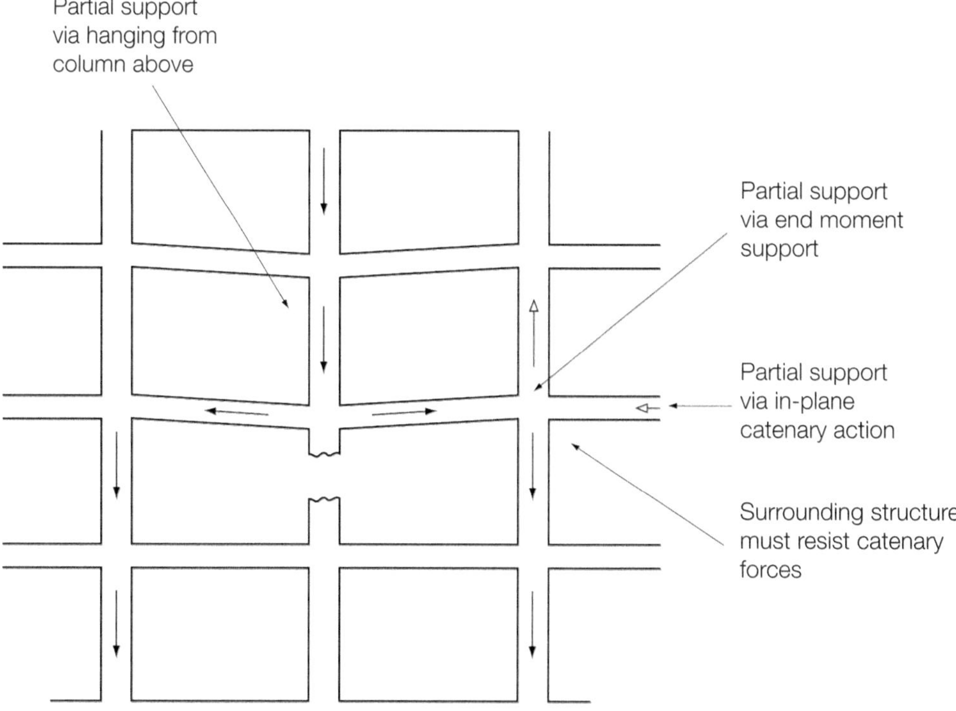

Currently there are no further formal guidelines available, although catenary action is currently being explored by European Project Team WG6.T2[50]. At present, catenary action is discussed for interest and to enable an engineer to carry out approximate calculations; it is unlikely that an engineer will undertake detailed calculations because of the complexity of modelling exactly what is occurring. In the model of a cable under a uniformly distributed load (UDL) in Figure 5.3, h (sag) is indeterminate and must be calculated by trial and error based on a maximum credible value of R_h and how much the beam at the support may be able to rotate.

Figure 5.3: Analysis of a tension member supporting a horizontally distributed load

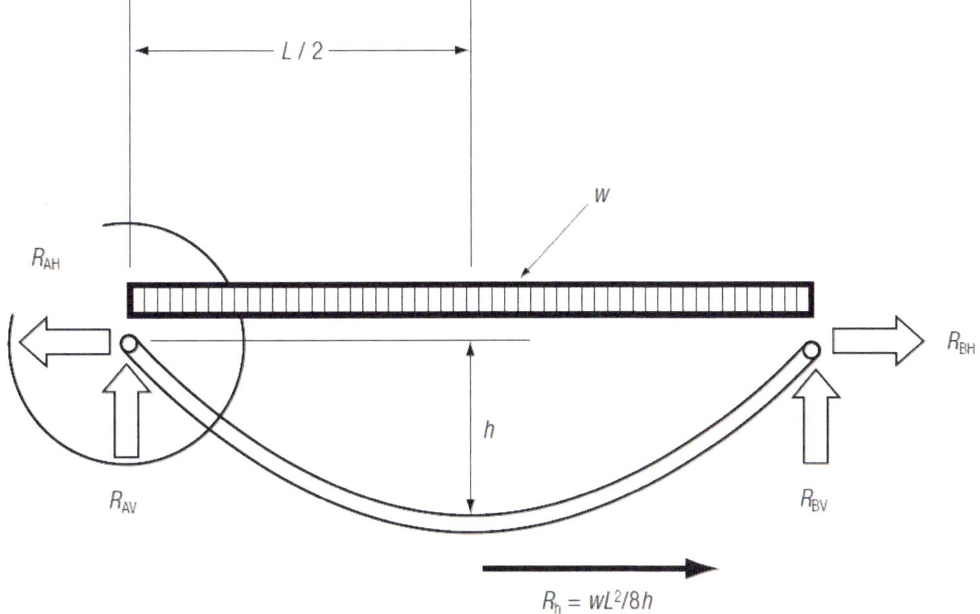

However, catenary action will be activated in a building if there is a failure of a structural element.

In two-way spanning structures, similar benefits can be achieved via radial tensile membrane action, and when considering the loss of an internal column below a slab (or equivalent) it is possible to develop an in-plane compression ring action to resist the radial tie forces generated.

Clearly catenary systems only work if there is enough ductility in the joints to form and sustain a mechanism, and if the catenary tension force can be anchored, the structure surrounding the tie end must also resist it. Where support from a column hanger is assumed, the structure around the hanger top must be capable of sustaining the additional loading and the beam capable of sustaining reversed shear.

 *Mandate M/515. Development of 2nd generation of EN Eurocodes. Robustness rules in material related Eurocode parts[50]*

5.6.2 Partial collapse and debris loading
Where the structure cannot support damage caused by the accidental event, limited collapse is permitted. Approved Document A stipulates that:

"The area of floor at any storey at risk of collapse does not exceed 15% of the floor area of that storey or 100m^2, whichever is smaller, and does not extend further than the immediate adjacent storeys."

This is clarified in BS EN 1991-1-7 as 'two adjacent storeys' (Figure 5.4).

Figure 5.4: Area at risk of collapse in the event of an accident

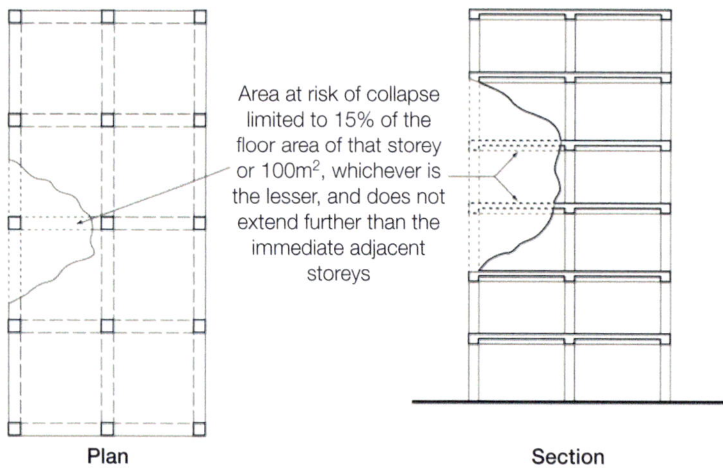

The limitation on storey spread implies that the floor below must be able to support debris accumulation.

Depending on the characteristics of the presumed collapse, the debris load may be less than the full load of the collapsed part. However, as the collapse of two adjacent floors is allowed, the debris loading may be that from two floors.

It is not possible to be specific and each case must be considered individually. However, an allowance for imposed load must be included, consistent with the accidental load case and with the most likely failure mechanism.

The Department for Communities and Local Government report[45] explores the concept of debris loading and it is covered further in Section 13.5.4.3 of this *Guidance*.

The 15% and 100m² limit is interesting because the consequence of this is that some structural forms and layouts will never be able to satisfy the notional element removal approach.

A masonry structure may have a room layout of approximately 4 x 5m. If it is assumed that a wall blows out, 20m² of the floor will collapse (Figure 5.5). Therefore, the floor area of that storey will need to be at least 20m² x 100/15 = 133m². This equates to approximately two two-bedroom flats, so is a relatively small building. The 20m² figure is at the lower end and assumes that there is a structural screed over precast planks or an *in situ* slab — if the floor were timber or precast without a screed, 40m² could be lost.

Figure 5.5: Area of floor lost if wall blown out

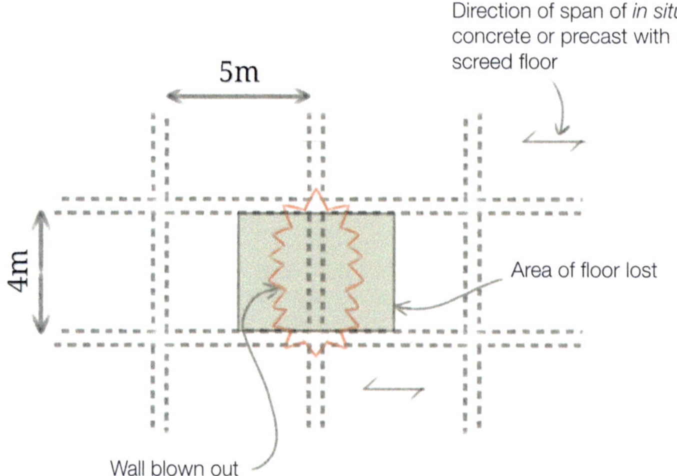

The steel structure in Case study 3 in Chapter 7 also looks at this, identifying that the building would lose over 200m^2 of floor, which is greater than the limit of 100m^2.

5.7 Approach 3: Key elements

A third approach, which is often easier to contemplate than notional element removal, is to consider members as key elements, in effect making them strong enough to withstand a prescribed hazard loading.

The approach is beneficial if such members are strong enough to survive. The key element approach offers advantages if failure of that single member would otherwise lead to loss of a significant portion of the building; examples include loss of a significant transfer beam or a column supporting such a beam.

The general design approach for key elements is to consider uniform pressure acting over their surface (in each orthogonal direction, one direction at a time) plus the surface of any attached items such as cladding.

Any member restraining a key element should be designed for the specified design pressure or the restraining element benefits ignored. The restraining element can be checked separately from the key element; however, if both would be affected by the same event, it is more logical to consider both elements loaded simultaneously. The pressure is applied in conjunction with the accidental load case.

A pressure value of 34kN/m^2 (derived from Ronan Point) is used as the test design pressure, representing the static equivalent pressure from a notional gas explosion. The force transferred from the attached item may be reduced to a realistic estimate governed by an upper bound of the fixing capacity.

It would be unduly onerous to apply the accidental design pressure over large areas, e.g., to slabs attached to (and stabilising) long span transfer beams. The 2.25h limit on length for loadbearing walls usually corresponds to no more than 6m, so applying the pressure to an area limited to 6m x 6m would seem reasonable, although any such relaxation should only be considered once specific circumstances have been factored in.

The pressure of 34kN/m^2 is onerous for walls and slabs but often non-critical for isolated columns due to their limited surface area. Structural sense should be used; the value of lateral load generated on an unusually narrow column may not be an appropriate design force. For all key elements, the possibility of other accident scenarios should always be considered.

Stuart Alexander[51] has proposed a static equivalent impact load of 150kN for columns, increasing to 250kN for the ground floor.

These values are reasonable for typical situations but again the actual load used should be justified on a case-by-case basis. For buildings adjacent to highways, further guidance on vehicle impact is given in BS EN 1991-1-7.

5.7.1 Transfer beams

The consequences of any transfer beam failure or of its supporting structure is likely to be more serious than the failure of standard beams or columns.

Where transfer beams are clearly carrying significant portions of a building, the standard tie forces could prove inadequate, and the recommended approach is to design the beams and associated structure:

- To sustain only localised damage, or
- To be removable elements, or
- To be designed as key elements

Transfer structures are very common in modern buildings, with several storeys of apartments with regular wall or column layouts built over retail, garaging or public areas with very different and wider column layouts. Therefore, design engineers must understand the impact on robustness of transfer structures, and will frequently be required to design podium floors and supporting columns as key elements, ensuring that they have capacity to sustain 34kN/m^2 and debris loading for any collapsed structure above (Figure 5.6).

Figure 5.6: Consideration of transfer structures

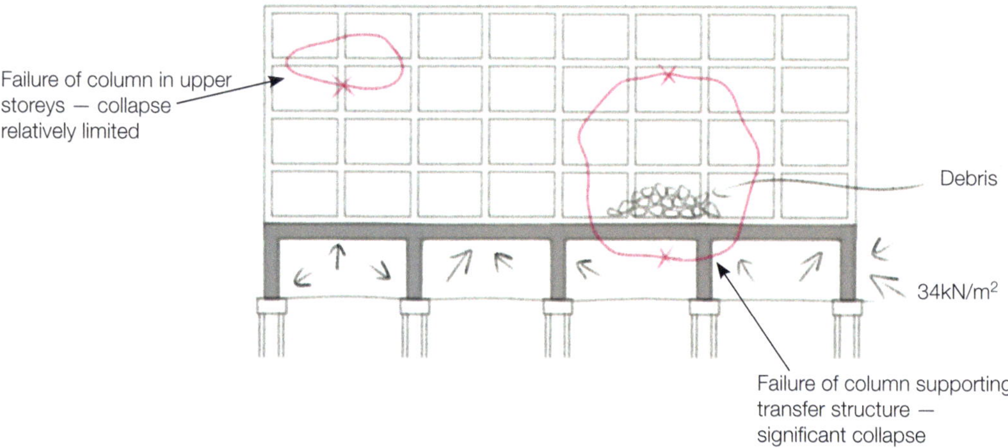

5.8 Design load cases

Calculations are required when assessing structures for imposed accidental loading, taking the notional element removal approach or undertaking key element design.

Three questions arise:

- What are the appropriate load factors?
- What loads should be used?
- Can material safety factors be reduced?

BS EN 1990 sets out the equation for the combination of actions for the accidental design situation in Expression 6.11b and in Table NA.A1.3 of the National Annex to the Eurocode (Figure 5.7):

Figure 5.7: Design values of actions for use in accidental and seismic combination of actions

Table NA.A1.3 — Design values of actions for use in accidental and seismic combinations of actions

Design situation	Permanent actions		Leading accidental or seismic action	Accompanying variable actions[b]	
	Unfavourable	Favourable		Main (if any)	Others
Accidental (Eq 6.11a/b)	$G_{kj,sup}$	$G_{kj,inf}$	A_d	$\Psi_{11}\,Q_{k,1}$	$\Psi_{2,i}\,Q_{k,i}$
Seismic[a] (Eq 6.12a/b)	$G_{kj,sup}$	$G_{kj,inf}$	$\gamma_1 A_{Ek}$ or A_{Ed}		$\Psi_{2,i}\,Q_{k,i}$

[a] The seismic design situation should be used only when specified by the client. See also Eurocode 8.
[b] Variable actions are those considered in Table NA.A1.1.

A simplified representation of Expression 6.11b, applicable to most structures is:

$$E_d = G_k + A_d + \Psi_{1,1}Q_{k,1} + \Sigma\,\Psi_{2,i}Q_{k,i}$$

Where:

E_d = design value of the effect of actions
G_k = characteristic value of a permanent action
A_d = design value of an accidental action
Ψ_1 = frequent value of a variable action
Q_k = characteristic value of a variable action
Ψ_2 = quasi-permanent value of a variable action

Combinations of actions for accidental design should either:

- Involve an explicit action, A_d
- Refer to a situation after an accidental event

When taking the notional element removal approach, the engineer would be looking at the performance of the building after the event and therefore A_d is 0.

When looking at key element design, A_d may be 34kN/m^2 or a specific accidental load such as vehicle impact.

The partial load factors for the accidental actions and variable actions, Ψ_1 and Ψ_2 depend on the building use and are found in Table NA.A1.1 of the National Annex to BS EN 1990 (Figure 5.8).

Figure 5.8: Values of Ψ factors in buildings

Table NA.A1.1 — Values of Ψ factors for buildings

Action	Ψ_0	Ψ_1	Ψ_2
Imposed loads in buildings, category (see EN 1991-1.1)			
Category A: domestic, residential areas	0,7	0,5	0,3
Category B: office areas	0,7	0,5	0,3
Category C: congregation areas	0,7	0,7	0,6
Category D: shopping areas	0,7	0,7	0,6
Category E: storage areas	1,0	0,9	0,8
Category F: traffic area, vehicle weight $\leq$ 30 kN	0,7	0,7	0,6
Category G: traffic area, 30 kN < vehicle weight $\leq$ 160 kN	0,7	0,5	0,3
Category H: roofs[a]	0,7	0	0
Snow loads on buildings (see EN 1991-3)			
— for sites located at altitude H > 1 000 m a.s.l.	0,70	0,50	0,20
— for sites located at altitude H $\leq$ 1 000 m a.s.l.	0,50	0,20	0
Wind loads on buildings (see EN 1991-1-4)	0,5	0,2	0
Temperature (non-fire) in buildings (see EN 1991-1-5)	0,6	0,5	0

[a] See also EN 1991-1-1: Clause 3.3.2 (1)

Table NA.A1.1 and NA.A1.3 are frequently used in calculating tie forces when using the prescriptive approach of providing ties when designing for robustness.

For some materials, partial material factors can be reduced when designing to the accidental load case, but this needs to be checked for each material.

5.9 Fire as an accidental load case

Within BS EN 1991-1-7, fire is an accidental load case. A fire is considered an accident that can:

- Affect the ability of a structure to sustain loads and transmit them to the ground safely
- Result in collapse disproportionate to the cause
- Challenge the ability of the structure to remain stable for a reasonable period with due consideration to the size and use of a building

These collective considerations fall under the term 'structural robustness in fire'.

Structural robustness in fire is covered in Chapter 14 and is intended to assist structural engineers when developing designs that must comply with the Building Regulations 2010[26] (as amended). Performance expectations for the structure of a building relevant to fire are contained in Requirements A1(1), A3 and B3(1) of Schedule 1 to the Building Regulations.

The chapter explains the basis and limitations of the simple 'fire resistance' design approach embedded within the approved documents and other guidance. It highlights some key implications and considerations for structural robustness in fire, and points to how structural engineers can and should engage with structural fire safety design more comprehensively. Signposts and references are provided to other standards and literature for further reading.

Case study 1

Example process of structural design

With the Building Safety Act expecting all buildings to be provided with a file detailing how the structure was designed, a document explaining the process with sketches and descriptions is important, and could also form a key part of the Design Intent Statement by the structural engineer.

The process of structural design could be:

1. Describe the structure and materials of the building, explaining the layout and loading.

2. Identify specific risks to the building, such as vehicle impact.

3. Identify areas of the structure that are particularly important, such as transfer beams or cantilevers, where failure would result in a large amount of the structure collapsing.

4. Determine whether the structure has a robust layout and identify where loadbearing elements do not align through storeys.

5. Decide which approach is going to be taken to ensure the structure is robust.

    5.1 If vertical and horizontal ties are to be provided, ensure that an alternative robustness strategy is employed if ties cannot be continuous in a straight line (Point 4)

    5.2 Calculate the horizontal and vertical tie forces. If there is an intersection of two materials (e.g., a steel frame supported on a concrete frame), calculate the tie forces for both materials and take a sensible view on which tie force should be used to connect the two materials together.

    5.3 If the notional element removal approach is taken, check the remaining structure above and ensure that the notional elements removed cover all possibilities (e.g., corner, perimeter and internal columns). Provide supporting calculations — hand calculations may be suitable for a masonry arch check, or a finite element analysis for a framed building. In addition, provide horizontal ties (and vertical ties if possible). Ensure collapse is limited to 100m^2 or 15%.

    5.4 Identify elements that need to be designed as key elements because their failure would cause greater collapse than that allowed.

6. Complete the supporting calculations.

7. Recheck the building to ensure that it is robust and that the three approaches have been used appropriately, and not solely as a mathematical exercise to get the building to 'pass'.

6 Classification of buildings

6.1 Introduction

One of the key aspects in the design of a robust building and the starting point of design is to determine the consequence class of the building. Consequence class (CC) is based on building type, number of storeys and occupancy. All these characteristics have an impact on the building's exposure to accidental actions and the consequence of those actions.

The type of accidental event that may occur in a two-story, single-occupancy house is very different from an accidental event that may occur in a hospital. The consequence of these events will depend on the layout and structure of the building, height, floor area and nature and number of occupants.

The consequence classes are defined both in Approved Document A[25] and BS EN 1991-1-7[4]. Figure 6.1 shows the consequence classes from Approved Document A.

Figure 6.1: Building consequence classes from Approved Document A

Table 11 **Building consequence classes**	
Consequence Classes	**Building type and occupancy**
1	Houses not exceeding 4 storeys
	Agricultural buildings
	Buildings into which people rarely go, provided no part of the building is closer to another building, or area where people do go, than a distance of 1.5 times the building height
2a Lower Risk Group	5 storey single occupancy houses
	Hotels not exceeding 4 storeys
	Flats, apartments and other residential buildings not exceeding 4 storeys
	Offices not exceeding 4 storeys
	Industrial buildings not exceeding 3 storeys
	Retailing premises not exceeding 3 storeys of less than 2000m² floor area in each storey
	Single-storey educational buildings
	All buildings not exceeding 2 storeys to which members of the public are admitted and which contain floor areas not exceeding 2000m² at each storey
2b Upper Risk Group	Hotels, blocks of flats, apartments and other residential buildings greater than 4 storeys but not exceeding 15 storeys
	Educational buildings greater than 1 storey but not exceeding 15 storeys
	Retailing premises greater than 3 storeys but not exceeding 15 storeys
	Hospitals not exceeding 3 storeys
	Offices greater than 4 storeys but not exceeding 15 storeys
	All buildings to which members of the public are admitted which contain floor areas exceeding 2000m² but less than 5000m² at each storey
	Car parking not exceeding 6 storeys
3	All buildings defined above as Consequence Class 2a and 2b that exceed the limits on area and/or number of storeys
	Grandstands accommodating more than 5000 spectators
	Buildings containing hazardous substances and/or processes

Notes:

1. For buildings intended for more than one type of use the Consequence Class should be that pertaining to the most onerous type.

2. In determining the number of storeys in a building, basement storeys may be excluded provided such basement storeys fulfil the robustness requirements of Consequence Class 2b buildings.

3. BS EN 1991-1-7:2006 with its UK National Annex also provides guidance that is comparable to Table 11.

There are some differences between Table 11 in Approved Document A and Table A.1 of BS EN 1991-1-7. Table A.1 states that a building to which large numbers of the public are admitted would be a CC3 building, and the area limit on a retail building up to three storeys is 1,000m^2.

This *Guidance* does not cover CC3 buildings, but it is important to be familiar with the lower consequence classes when designing a CC3 building. It is also vital to remember the key principle of designing a CC3 building when working with CC1, CC2a and CC2b buildings — that of identifying risks.

There will always be buildings not covered by Tables 11 and A.1 which may need to be treated as a CC3 building, because they either house high-risk activities or because the consequence of failure would be serious. Examples might include:

- Data centres
- Industry control centres
- Significant religious buildings
- Museums and galleries
- Airports and transport hubs
- Large innovative structures

6.2 Number of storeys

The number of storeys in a building is key to its classification. However, what counts as a storey is sometimes less obvious.

6.2.1 Basements

Both Approved Document A and BS EN 1991-1-7 are clear that if basements are designed to Consequence Class 2b, they do not need to be included in the storey height (Figure 6.2). The logic of this decision is obscure, but may derive from the inherent robustness of a basement. Basements should only be treated in this way if all four walls are present. Part basements should be treated as an above-ground storey.

Figure 6.2: Consideration of basements in consequence classes

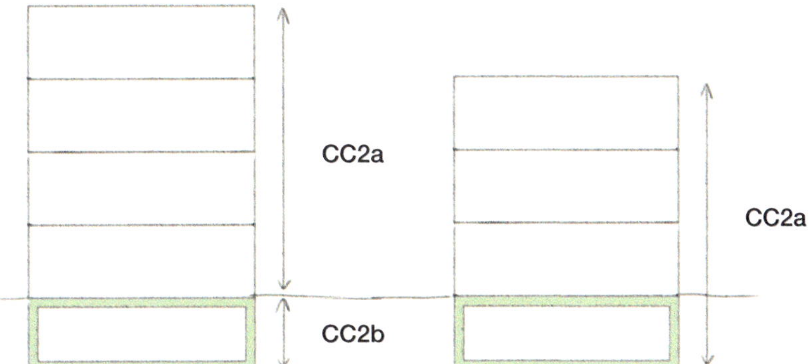

6.2.2 Warranty providers

Major warranty providers, such as NHBC, LABC Warranty and Premier Guarantee extended this 'storey exclusion' to certain five-storey buildings. NHBC issued guidance notes allowing a five-storey residential building to be designed as CC2a, if the ground floor storey was designed as CC2b and with the structural elements designed as key elements[52].

The floor forming the podium would be identified as a key element and would need to be designed to carry the debris from the collapse and the accidental load of 34kN/m^2. The columns supporting the structure would also be designed as key elements. Although the fall of debris is clearly a dynamic event, in the method of design adopted, the presumed loading was simply the self-weight of the debris.

This approach was commonly accepted by the warranty providers in the UK, and therefore has been widely adopted by consulting structural engineers.

The Department for Communities and Local Government report[45] expressed a number of concerns about this approach and believe it to be:

"Unconservative and particularly ill-advised."

It is agreed that this is an over-generous interpretation of Table 11 of Approved Document A.

In light of the risks set out in the DCLG report, it has been accepted that the approach set out by NHBC is no longer advisable, as design engineers are required to take account of up-to-date guidance on risk.

Therefore, it is not advised that the ground-floor storey should be excluded; a five-storey residential building should always be classed as CC2b, throughout each storey.

6.2.3 Attics, roofs, tall storeys, mezzanines and basements

A very common issue encountered is when an attic storey is only partially inhabited. It is difficult to know if the attic should be counted as a storey or if it can be ignored. A design engineer will have to make a judgement on the risk and significance of a small attic storey, and at what point the attic space is sufficiently inhabited for it to be considered as a storey. The DCLG report recommends that habitable roof space should be counted as a storey, regardless of the roof slope.

The design engineer has the same problem when there is a small fifth storey as part of a residential building. Can the building be designed as Consequence Class 2a or does the small fifth storey tip the building into Consequence Class 2b?

A similar uncertainty occurs over whether a mezzanine or gallery counts as a storey. The decision may depend on its area and use. For mezzanine floors it is recommended that each situation should be judged individually, but an approximate guide is offered that a mezzanine floor should only be considered as a storey if it is greater than 20% of the building footprint[44].

To qualify as a basement storey, a basement should be deeper than 1.2m and greater than 50% of the plan area of the building[45].

Storey height should also be considered. It has been known for an industrial building to be labelled as two-storey, and appear to be a low-rise building, but it is later revealed that each storey is 10m high and that one end of the building area has six internal storeys of office accommodation.

For buildings with a varying number of storeys that fall into more than one class, guidance recommends that the robustness measures for the more onerous class should continue until a structural discontinuity is reached. This may be in the form of a party wall or similar natural architectural segregation within the structure. The engineer should be wary of treating a movement joint as a structural discontinuity because the movement joint may not delineate a building in the way it is used and occupied (Section 6.3.3).

It is important that design engineers use their judgement to assess the number of storeys and provide the rationale for that decision.

6.3 Practical problems of interpretation

Although the categories are reasonably precise, they should be interpreted with common sense and careful judgement.

For example, having a floor area of 2,001m^2 does not automatically imply upwards classification. Conversely, having a floor area of 1,999m^2 requires care in considering classification.

6.3.1 Multiple uses

The footnote to Table 11 of Approved Document A states:

"For buildings intended for more than one type of use the consequence class should be that pertaining to the most onerous type."

Typical situations falling into this category are apartments located over a part of the structure with another use, such as parking, shops (which include restaurants), or a nursery. Another is where offices are combined with apartments or hotels.

However, offices, hotels, apartments and other residential buildings all have the same cut-off levels, i.e., Consequence Class 2a up to four storeys and Consequence Class 2b up to 15 storeys, so there is no problem when determining consequence class.

With regard to shops, the Consequence Class 2a limit comes down to three storeys but the Consequence Class 2b limit remains unchanged. If a nursery is treated as educational, then any building of more than one storey will fall into Consequence Class 2b.

6.3.2 Undefined uses
There are many building uses which are not explicitly covered by Table 11 of Approved Document A.

Nurseries and kindergartens probably need to be treated as educational, as the number of children per unit area is consistent. Consequence class is more difficult to determine for doctors' surgeries and other day surgery units, care homes, hospices and other healthcare facilities where the number of people inside varies at any one time.

The appropriate consideration is probably whether people are sleeping on the premises, and if they are bedridden or vulnerable. It is logical to treat doctors' surgeries and day surgery units as offices, care homes as residential, depending on the vulnerability of the residents, but hospices as hospitals.

Approved Document A refers to two reports which discuss the evaluation of risk and consequence and these should be considered in unusual cases.

6.3.3 Building or structure?
Approved Document A refers to building classes, and uses the word 'building' throughout. BS EN 1991-1-7 also uses 'building' throughout, but replaces the Approved Document A term 'grandstand' with 'stadium'.

It is very tempting to design a building separated by a movement joint as two separate structures, especially if one side of the movement joint could be considered as CC2a and the other as CC2b, and SCI Publication P391[53] includes a diagram showing that a structure separated by a movement joint can be considered as two separate buildings. Some design engineers have taken this decision many times, because a Consequence Class 2a building can be designed less onerously and so save the developer money.

However, the wording of both Approved Document A and BS EN 1991-1-7 suggests that this is not what is intended (i.e., for downgrading classification) especially where an incident could affect both sides of the joint, and this view is reinforced by the example of the Paris airport terminal collapse (Case study 2 and Section 6.2.3).

This principle should not be applied too rigidly. In a hospital with a number of single-storey ward blocks grouped around a four-storey block, it would be reasonable to treat each block as a separate building, provided that each block was structurally independent and robust in its own right.

6.3.4 Acoustics
A particular problem arises in taller blocks of flats which need to be designed for both sound insulation and robustness, and the requirements for each are conflicting. For reasons of acoustic isolation, tying of units between flats is not normally permitted. This is an area of difficulty for which little guidance is available. Until further guidance is produced, it is recommended that advice is sought from unit suppliers who may have specific test data available. Acoustic separation is important and covered in legislation, but the structure cannot be compromised.

6.3.5 Strong floors
Strong floor design was referenced in the first edition of this *Guidance*. It stated that:

"If a strong floor can be designed to withstand collapse of the structure above, it clearly protects the occupants below. So a strong floor can be considered to be the foundation for the floors above and the number of storeys counted from this level up. However, below the strong floor the risk is unchanged, and the storey count should be that of the whole building. An example of the advantage of strong floors is the partial collapse of Torre Windsor in Madrid[54]."

Strong floor design of this type is rarely seen in the UK and it is not mentioned in Approved Document A and BS EN 1991-1-7.

However, the NHBC's 2004 document on robustness[52] included a building classification option that used the principles of strong floor design; a five-storey residential building with a first-floor podium deck and four storeys of timber frame, LSF, CLT or masonry construction over. Concerns expressed in the DCLG report should be noted.

Designing a tall building with strong floors at every few storeys is outside the scope of this *Guidance* and the engineer should look for other information and ensure that all risks are assessed, e.g., if one of the strong floors or its supports were to fail.

6.3.6 Modular and temporary buildings

There are building types which in the past have been used as temporary structures but are now being adopted as permanent buildings. An example of this is modular buildings which have traditionally been used as temporary classrooms in schools, but are now being erected to be permanent structures. Any structure that is intended to have a design working life of more than a few years should be designed for robustness and disproportionate collapse, even if there has not been a tradition of thinking of this building type in such a way.

This is not to say that good design can be ignored for temporary structures. For example, temporary grandstands may only be used for a few days, but they must still be robust. There have been catastrophic failures of such stands due to a lack of robustness. The IStructE's Advisory Group on Temporary Structures (AGOTS) has published guidance[55].

Case study 2

Paris Charles de Gaulle Airport, 2004 terminal collapse

Construction of the terminal was started by Aéroports de Paris in 1999 and was opened to the public in June 2003. It comprises a main building, a boarding jetty, and a link between the two. The jetty is 650m long and is made up of a succession of ten concrete shells. On 23 May 2004, at 6:57am, the roof of the boarding jetty collapsed over a length of 30m, killing four people and injuring seven (Figure 6.3). The toll could have been much higher if two police officers, who noticed cracking, had not evacuated waiting passengers a few minutes before the collapse.

Figure 6.3: Paris Charles de Gaulle Airport terminal collapse

The 300mm thick roof shells were elliptical, or horseshoe shaped, in section and strengthened externally by steel hoop-shaped frames, connected to the outside face by steel struts bearing on end-plates embedded near the surface of the concrete. From the time of construction until the failure event, cracking in various locations had been noted, and in some case remedial action had taken place. The police officers who had noted distress in the structure saw spalling on the inside concrete face around one of the 500mm square plates carrying compression from a strut.

There were several investigations with the conclusion that the trigger was a punching shear failure of the shell by the struts, as a consequence of inadequate shear resistance of the reinforced concrete. The safety factors in the structure were deemed to have been very low from the outset, and it was thought to have been weakened by creep, shrinkage, temperature effects and wind[56].

After the punching failure of the end-plate pushing through the shell, there was a redistribution of forces in an already-cracked structure, and this led to the bending failure of one of the solid arch elements of the shell due to the development of a plastic hinge. This rupture was followed immediately by the rupture of other parts of the structure and hence the collapse.

The lessons learnt were that insufficient attention had been paid to the complex nature of this unusual structure, resulting in a lack of redundancy and inadequate consideration of punching shear requirements. In short, there was a lack of robustness in the structure.

7 Design of steel buildings

7.1 Introduction to steel design

Steel is used for a variety of buildings but most commonly in multi-storey buildings or portal frames used for retail, industry or agriculture. Resistance to wind and other lateral loads is generally provided through bracing or portalised frames.

Structural steel elements are also used in hybrid structures such as masonry and timber-framed buildings, where larger openings are required, or loading and spans are such that timber beams or masonry arches are insufficient.

Also common are low-rise masonry buildings with precast plank floors, steel beams to support the planks and a trussed rafter roof. In city centres and on large new housing developments, multi-storey steel frames placed on top of a concrete podium structure are a popular building form.

Although the design of structural steel is relatively simple, the engineer also needs to be aware of the other materials used in the structure, and ensure that the rules are satisfied for each material and that the interfaces of different materials have been considered.

Although structural steel buildings are inherently robust — and the prescriptive tying rules are straightforward to calculate and typically implemented through standard connection design — it is necessary to be aware of higher-risk structural forms requiring special consideration such as transfer beams. Engineers should also consider the impact of an accident on any stability system, remembering that bracing is frequently pin-jointed and a single bracing system will not provide structural redundancy.

This chapter follows the requirements of BS EN 1993-1-1[36] and BS EN 1991-1-7[4] and the supplementary National Annexes[57,33]. Eurocode 3 contains no specific requirements for robustness so it is necessary to refer to Annex A in BS EN 1991-1-7 for guidance.

Another key reference is the SCI Publication P391[53] to be read in conjunction with the SCI Advisory Desk publication AD-415[58].

7.2 Summary of requirements for each consequence class

If CC1 buildings are designed to Eurocode 3, no additional measures are required. It is good practice that CC1 buildings (of which there will be some) follow the tie requirements of CC2a.

CC2a buildings should be provided with horizontal ties for framed construction or effective anchorage of suspended floors to walls for loadbearing wall construction, as defined in A.5.1 or A.5.2 of BS EN 1991-1-7.

For CC2b buildings, horizontal ties, as defined in A.5.1 for framed buildings and A.5.2 for loadbearing wall construction, together with vertical ties, as defined in A.6 in all supporting columns and walls, should be provided.

Alternatively for CC2b buildings, the notional element removal or key element approach can be taken.

CC2b buildings will probably require horizontal ties to satisfy the requirements of the notional element removal approach. This approach is not commonly applied to steel-framed structures because of the ease and simplicity of providing horizontal ties.

7.3 Horizontal loads

BS EN 1993-1-1 specifies that, in addition to wind load, allowance must be made for imperfections. These include residual stresses as well as geometrical imperfection. Global imperfections for frames and bracing systems must be taken into account in the analysis, and this is usually done by converting the imperfections to equivalent horizontal forces. Second-order effects may also need to be considered, even on low-rise buildings.

7.4 Tying

A.5 (horizontal ties) and A.6 (vertical ties) in Annex A of BS EN 1991-1-7 include the formulae required to calculate the tie forces. BS EN 1993-1-8[59] does not provide any specific guidance on the tying resistance of connections for robustness purposes.

7.4.1 Provision of ties
Three types of ties are specified:

- Peripheral ties
- Internal ties
- Vertical ties

Large strains and large deformations are acceptable in the accidental design situation. Therefore, for the calculation of connection tying resistance, SN015[60] recommends that ultimate tensile strengths (f_u) be used and the partial factor for tying $\gamma_{M,u}$ be taken as 1.1[61].

Refer to SCI P364[62] and SCI P358[61] for examples of connection design.

7.4.2 Horizontal ties
Horizontal tie forces are calculated using Expressions A.1 and A.2[4].

For internal ties $T_i = 0.8(g_k + \Psi q_k)\, s\, L$ or 75kN, whichever is greater
For perimeter ties $T_p = 0.4(g_k + \Psi q_k)\, s\, L$ or 75kN, whichever is greater

Where:
s = spacing of ties
L = span of the tie
Ψ = relevant factor in the expression for combination of action effects for the accidental design situation (i.e., Ψ_1 or Ψ_2 in accordance with Expression 6.11b of BS EN 1990[34])
g_k = permanent action
q_k = variable action

Expression 6.11b: $\sum_{j\geq1} G_{k,j}"+"P"+"A_d"+"(\Psi_{1,1}\ \text{or}\ \Psi_{2,1})Q_{k,1}"+"\sum_{i>1} \Psi_{2,i}Q_{k,i}$

Where P is a prestressing action[34].

When using A.1 and A.2 the value of Ψ is $\Psi_{1,1}$ applied to a variable load depending on the use of the building.

In the case of an office building, the engineer would choose a Ψ of 0.5 and a variable load of 3kN/m^2. If it were a lecture theatre, the engineer would choose a Ψ of 0.7 and a variable load of 4kN/m^2.

For values of $\Psi_{1,1}$, the design engineer should refer to Table NA.A.1 of the National Annex to BS EN 1990[44] and Section 5.8 of this *Guidance*.

A.5.1 of BS EN 1991-1-7 states that horizontal ties should be provided:

- Around the perimeter of each floor and roof level
- Internally in two right-angle directions to tie column and wall elements securely

The horizontal ties should be continuous and arranged as closely as practical to the edges of floors and lines of columns and walls. At least 30% of the ties should be located close to the grid lines of the columns and walls.

What this means practically, is that primary and secondary beams are most likely used as the horizontal ties, along with the perimeter beams. Additional horizontal ties can be provided by the reinforcement in floors made of concrete and profiled metal sheets.

BS 5950-1[63] (withdrawn) required heavy floor units to be anchored to supporting beams, or to each other over a supporting beam. This is not a requirement in BS EN 1991-1-7 but remains a recommendation of the Steel Construction Institute (SCI P391).

7.4.3 Vertical ties

A.6 (1) and (2) of BS EN 1991-1-7 state:

"Each column and wall should be tied continuously from the foundation to the roof level.

In the case of framed buildings the columns and walls carrying vertical actions should be capable of resisting an accidental design tensile force equal to the largest design vertical permanent and variable load reaction applied to the column from any one storey.

Such accidental design loading should not be assumed to act simultaneously with permanent and variable actions that may be acting on the structure."

There are four points of confusion for designers for the calculation of vertical ties, because the approach is different in BS EN 1991-1-7 from that in BS 5950. These are:

1) The phrase *"…largest design vertical permanent and variable load reaction applied to the column from any one storey"*
2) What are the permanent and variable loads, and how are they factored?
3) What is meant by accidental load?
4) What happens to the derived vertical tie force at foundation level?

BS 5950-1, Clause 2.4.5.3 (c) stated:

"All column splices should be capable of resisting a tensile force equal to the largest total factored vertical dead and imposed load applied to the column at a single floor level located between that column splice and the next column splice down."

In Eurocode designs, the entire column length (and any splice) should be capable of carrying the largest accidental design tension resulting from any one storey. However, in BS 5950-1 design, the column splice and column only needed to be designed for the vertical loads coming in at the storey above the column splice.

The impact of this difference in codes would be seen if there was a floor with particularly large loading (perhaps because of larger spans or transfer structures). The vertical tie force derived from this floor would be greater than the tie force derived from other floors, but would need to be applied throughout each column and splice if the building was designed to Eurocodes.

The loads that are chosen to calculate the tie force are the permanent and variable loads, g_k and q_k. The value of Ψ is $\Psi_{1,1}$ from Table NA.A.1 of the National Annex to BS EN 1990 and Section 5.8 of this *Guidance*.

Therefore, in the case of an office building, the vertical tie force would be:

$$T_v = (G_k + 0.5 \times Q_k) \times s \times L$$

The confusion identified in point 3) is that although the phrase 'accidental design load' is written in the code, an accidental load is not used in the calculation of the tie force; it is 'normal' permanent and variable loads that are put into the calculation. With respect to 3), it is another idiosyncrasy where, despite this being about accidents, the accidental load ($A_d = 34kN/m^2$) is not included although the load factors are taken from the accidental load case.

Expression 6.11b of BS EN 1990 can be used in either of two ways:

- Involving an explicit accidental action A (fire or impact), or
- Referring to a situation after an accidental event, where $A = 0$

Within this chapter, the latter is used in the calculation of horizontal and vertical tie forces because it is being used with the 'prescriptive rules' associated with the strategy where risks are unidentifiable.

The confusion in point 4) arises because A.6 (1) seems ambiguous to many engineers, who try and take the derived vertical force into the foundation. This is not what is expected; all that is needed is that the tie (effectively the column) should start at foundation level and end at roof level. It is advantageous that the tie force does not need to be transferred into the foundation, because these forces can often be hundreds of kN and would require significant strength from the holding-down bolts.

However, in the case of a steel structure on top of a reinforced concrete podium, the direct connection of roof to foundation is interrupted by the reinforced concrete podium. If the columns of the steel and reinforced concrete elements align, vertical tie forces could be calculated for both steel and concrete and the maximum tie force value used throughout the building, with the steel column base fixing designed to transfer the tie force from column to podium deck.

Note that the calculation of vertical tie forces in a reinforced concrete structure is subtly different from that in a steel frame because tie forces for reinforced concrete are not derived from BS EN 1991-1-7 but from Eurocode 2 (see also Section 9.3.5 of this *Guidance*).

If the reinforced concrete podium structure is to act as a transfer structure, with a greater column spacing at ground floor than that of the steel frame upper floors, a different approach would need to be taken because the columns, and consequently the vertical ties, will not align and ties should run in a straight line (Sections 5.5.1 and 7.8.2).

7.5 Notional element removal and key element design

An alternative to following the prescriptive rules for tying is to consider notional element removal and to investigate the subsequent collapse area, following the principles detailed in Chapter 5. There are no special requirements for steel structures, except that partial material factors and partial load factors can be reduced to those in the accidental load case taken from Table NA.A.1 of the National Annex to BS EN 1990. These are discussed in Sections 5.8 and 7.4.1 of this *Guidance*.

Key element design is rarely the approach taken when designing steel structures, because it is generally straightforward to achieve the horizontal and vertical tie forces with standard steel connections.

However, if there is the possibility of accidental loads, such as vehicle impact loads in ground-floor car parks or fire engines returning to a fire station through an entrance framed by key support columns, these need to be designed for[64].

Designing for accidental loads is often used in conjunction with the prescriptive approach of providing horizontal and vertical ties.

7.6 Bracing systems

A key to all multi-storey frames is bracing, which is generally in the form of cross-bracing with pinned connections to the frame. Pinned structures have no redundancy, so it is important not to rely on just one bay of bracing. Usual practice is to provide two or more braced bays in each direction.

Each floor should be connected to the bracing system. This may not be possible if there are some double-height spaces adjacent to the braced bay. In this case the structural engineer may need to consider what would happen if there were a loss of supporting structure in this area of the building.

If a braced bay were to be lost, any check of the remaining structure does not need to use the factored loading, but the accidental loading described in Expression 6.11b of BS EN 1990 which can be simplified to:

$$G_k + \Psi_{1,1} \times Q_{k,1} + \Psi_{2,2} \times Q_{k,2}$$

7.7 Floor units

Steel usually provides only the frame skeleton, the floors being either *in situ* concrete on metal decking (which may also act as a horizontal diaphragm) or precast concrete units.

There are no specific requirements in BS EN 1991-1-7 regarding the anchorage of floor units.

There were provisions in BS 5950-1 regarding anchorage of floor units down to the frame, with the aim of preventing the floor falling through the frame if the steelwork is moved, or the floor units uplifted as a result of accidental loading (e.g., explosion).

BS 5950-1, Clause 2.4.5.3 (e) required that stairs, precast concrete or other heavy floor or roof units were effectively anchored in the direction of their span, either to each other over a support, or directly to their supports. Anchorage was only required in the direction of the unit span, as the steel beams act as ties in the orthogonal direction. The provisions of BS 5950-1 should continue to be followed.

SCI P391 also recommends that where precast concrete or other heavy floor, stair or roof units are used, they should be effectively anchored in the direction of their span, either to each other over a support or directly to their supports.

Tying of the floor units to the beams will often be necessary for purposes other than reducing sensitivity to disproportionate collapse, such as to mobilise floor diaphragm action against wind loading.

Eurocode 2 does not cover anchorage of precast floor and roof units and stair members. However, PD 6687[65] advises that the same requirements as given in BS 8110-1[39] should be used. Precast floor, roof and stair members should be effectively anchored, irrespective of whether such members are used to provide other ties required by BS EN 1992-1-1[37], Clause 9.10.2.

SCI P391[53] provides detailed guidance on anchorage of precast units across a range of support configurations.

Careful planning is required to ensure that planks are matched with steel beams on-site, especially if ties are pre-welded onto steel beams and because planks are precambered.

It is also increasingly common that cross-laminated timber (CLT) is used as floors and walls, in conjunction with steel frames. It must be confirmed that the structure satisfies the requirements of both the timber and steel codes, and CLT floor units should be anchored to the steel frame.

7.8 Consideration of structural form

7.8.1 Multi-storey braced frames
A multi-storey steel frame with repeated column arrangements at each level and bracing is inherently robust. The challenges for the structural engineer occur when designing a hybrid structure or a portal that has several internal storeys.

7.8.2 Steel buildings over podiums

If the building is structured with reinforced concrete columns at ground floor and a concrete podium deck (e.g., to allow pedestrian movement or car parking) with a multi-storey steel structure over, the engineer will need to consider the interface between the steel and concrete, and ensure that robustness has been resolved for both materials (Figure 7.1).

Figure 7.1: Steel frame over reinforced concrete podium

As the podium structure is generally a transfer structure, to enable a much more open space at ground floor level, using the columns as vertical ties from roof to foundation is not viable because of the lack of continuity.

One solution would be to design the podium structure for the vertical tie forces, which would put the podium into shear and bending. However, the design route generally taken is to design the supporting elements (in this case the columns and podium structure) as key elements for the accidental load and also debris loading.

SCI P391 includes worked examples with steel transfer structures at first-floor level, and includes both the horizontal and vertical tie approach and the horizontal tie and key element design approach.

7.8.3 Portal frames

If a portal frame were to be used for an agricultural building it would fall into Consequence Class 1. However, if it were to be used for a very large retail unit, with a floor area greater than 1,000m^2 it would be classed as a CC2b building. If the building was a simple warehouse but with a three-storey office at one end, it would be a CC2a building.

It is increasingly common to convert agricultural sheds into retail or educational facilities (farm shops, home school hubs, etc.) and an engineer would need to check whether there will be a change in building class and if so, that the existing structure satisfies the requirements of the new building class (Figure 7.2).

Figure 7.2: Agricultural building conversion

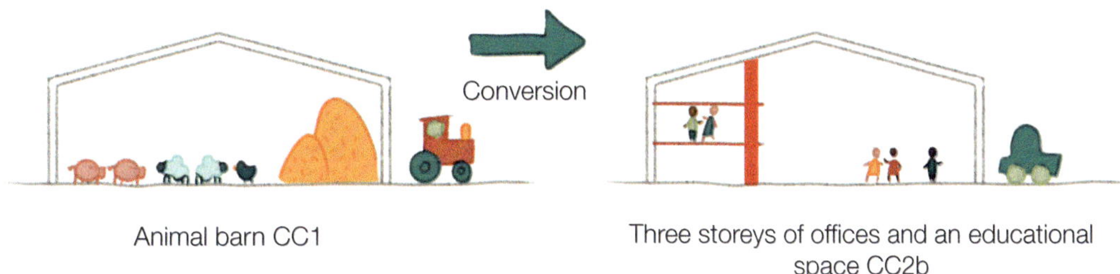

If the existing structure does not have capacity for the new building class, connections may need to be strengthened or additional structure introduced. It is likely, though, that because deflection limits are more onerous for a human-inhabited building than an animal shed, structure will be added and there is plenty of opportunity to provide the additional capacity required for the change in building consequence class.

7.9 Connection design

Connections must be designed to provide the required tie resistance, but they should also be ductile so that energy due to an accidental event can be absorbed and forces can be redistributed.

Connections should be designed for minimum force levels, at a size proportional to the elements they are connected to, and welds that are sufficient in size. A robust connection detail should be such that inherent assumptions about load redistribution can be realised.

Industry standard connections will generally comply with robustness objectives for conventional buildings. For more unusual structures, greater consideration of how the connection may perform when severely distorted may be required.

Case study 3

13-storey, mixed-use steel building — tying

The building is a 13-storey building with braced cores for stability.

The ground floor is used as retail space and its height is 5m. The first floor is office space with a height of 4m, and all other floors are residential with a height of 3m.

Figure 7.3: Section through 13-storey building

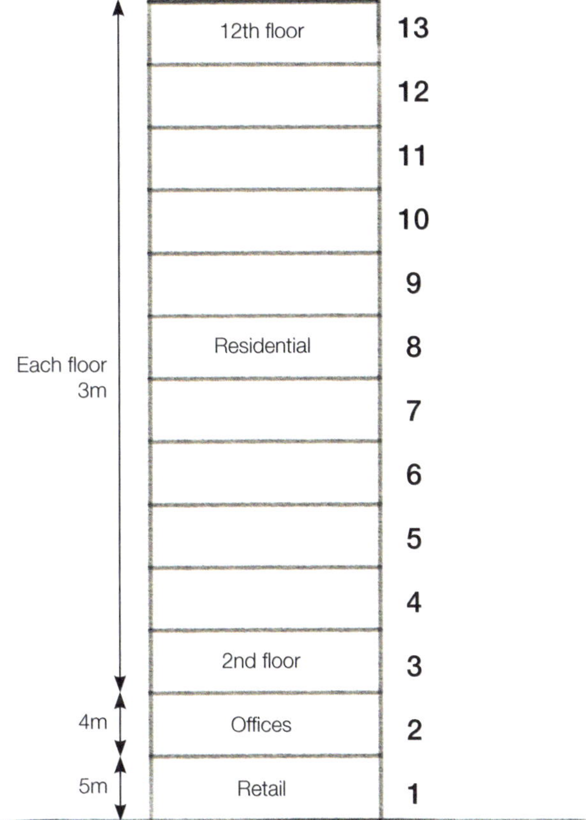

The columns are laid out on a 6m x 9m grid with the primary beams spanning 6m and the secondary beams spanning 9m. The spacing of the secondary beams is 3m.

A composite flooring system is used with steel decking spanning between the secondary beams. All secondary and primary beams are assumed to act composphericallyite with the floor slab. The steel frame has braced stair cores (Br) for stability (Figure 7.4).

Figure 7.4: Plan showing layout, column references and braced core

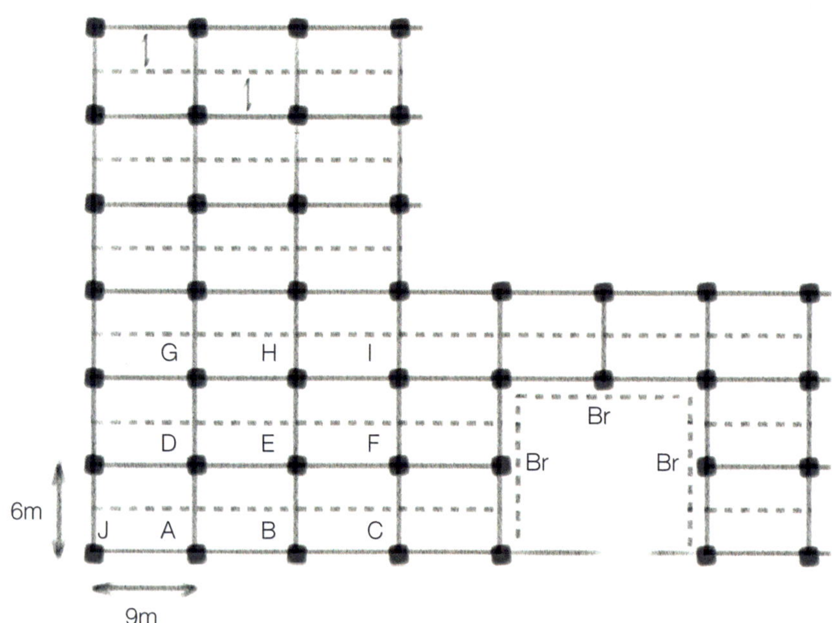

The composite floor system comprises steel decking spanning between the secondary beams with a 125mm-thick slab in grade C25/30 concrete.

Loading

The floor loading is:

Permanent action $\qquad$ $g_k = 3.5 \text{kN/m}^2$

Variable action, ground floor $\qquad$ $q_k = 7 \text{kN/m}^2$
Variable action, first floor $\qquad$ $q_k = 5 \text{kN/m}^2$ (including a partition allowance)
Variable action, 2nd–12th floor $\qquad$ $q_k = 2.5 \text{kN/m}^2$ (including a partition allowance)
Variable action, roof $\qquad$ $q_k = 4 \text{kN/m}^2$

The permanent action due to cladding is:

Rainscreen cladding $\qquad$ $g_k = 1.5 \text{kN/m}^2$

Building classification

There is a mix of uses within the building, but at 13 storeys the consequence class is the same for retail, office and residential and is CC2b.

Applying the tying method for robustness

Horizontal ties

$T_i = 0.8(g_k + \Psi q_k) s L$ or 75kN, whichever is greater

$T_p = 0.4(g_k + \Psi q_k) s L$ or 75kN, whichever is greater

For both residential and office spaces $\Psi = 0.5$. For retail $\Psi = 0.7$

The primary beams and secondary beams act as the internal ties.

On the residential floors
T_i (primary beam) = 0.8 x (3.5kN/m^2 + 0.5 x 2.5kN/m^2) x 6m x 9m = 205kN

T_i (secondary beam) = 0.8 x (3.5kN/m^2 + 0.5 x 2.5kN/m^2) x 3m x 9m = 103kN

The perimeter tie forces (T_p) will be half these internal tie force values.

On the office floor
T_i (primary beam) = 0.8 x (3.5kN/m^2 + 0.5 x 5kN/m^2) x 6m x 9m = 260kN

T_i (secondary beam) = 0.8 x (3.5kN/m^2 + 0.5 x 5kN/m^2) x 3m x 9m = 130kN

The perimeter tie forces (T_p) will be half these internal tie force values.

Vertical ties
The entire column length (and any splice) should be capable of carrying the largest accidental design tension resulting from any one storey. The office storey has a greater imposed load, therefore it is this storey that should be used for calculation of the vertical tie force.

Internal columns, D, E, F, G, H and I
T_v = (3.5kN/m^2 + 0.5 x 5kN/m^2) x 6m x 9m = 324kN

Perimeter columns, A, B, C
The vertical tie force includes cladding load for a storey height (on assumption that cladding is supported at every storey).

T_v = ([3.5kN/m^2 + 0.5 x 5kN/m^2] x 3m x 9m) + (1.5kN/m^2 x 4m x 9m)

   = 162kN + 54kN

   = 216kN

Corner column, J
The vertical tie force includes cladding load.

T_v = ([3.5kN/m^2 + 0.5 x 5kN/m^2] x 3m x 4.5m) + (1.5kN/m^2 x 4m x [4.5m + 3m])

   = 81kN + 45kN

   = 126kN

An alternative to providing vertical ties could be to provide horizontal ties in conjunction with checking the notional removal of a structural element.

If column E of this building were to be removed, the floor would be lost between columns A, G, I and C. This represents an area of 18m x 12m = 216m^2 and is greater than the 100m^2 allowed. It is also likely that some floor area at higher levels would be lost.

Therefore, the element removal approach is not appropriate for a steel building with this structure and layout.

Additional structural provisions
The recommended additional structural provisions relate to:

- Vertical bracing
- Anchorage of heavy units

Vertical bracing

The braced bays should be distributed throughout the building such that, in each of two directions approximately at right angles, no substantial portion of the building is connected to only one system for resisting horizontal force.

Anchorage of heavy units

The floor and roof slabs should be effectively anchored in the direction of their span, either to adjacent slabs over a support, or directly to their supports.

The anchorage should be capable of carrying the self-weight of the floor or roof unit.

The composite floor slabs span 3m between supports.

It is assumed that their dead weight is equal to the floor permanent action, $g_k = 3.5kN/m^2$.

Required anchorage = $3.5kN/m^2 \times 3m/2 = 5.25kN/m$ width.

Using a properly anchored A142 mesh reinforcement (cross-section area of $142mm^2/m$ and yield strength of $500N/mm^2$) will provide a tensile anchorage resistance of 71kN/m width over a support.

Note: For robustness, the material partial factor for the accidental design situation may be used, which is 1.0 for reinforcing steel.

For the slabs spanning onto edge beams, a suitable detail will be needed to anchor the floor slab directly to the edge beam.

If edge beams are designed as composite with the use of welded shear studs, the anchorage requirements will generally be satisfied.

If edge beams are designed as non-composite, welded shear studs may be used to provide the required anchorage, even though the edge beam is non-composite.

Alternatively, the shot-fired pins that are used to connect the deck to the supporting steelwork may be used. The resistance of a shot-fired pin connection is dependent on the fixing used and decking thickness. For a shot-fired pin connection with a resistance of 2.7kN, sufficient anchorage is achieved for this example with shot-fired pins spaced at 500mm along the beam.

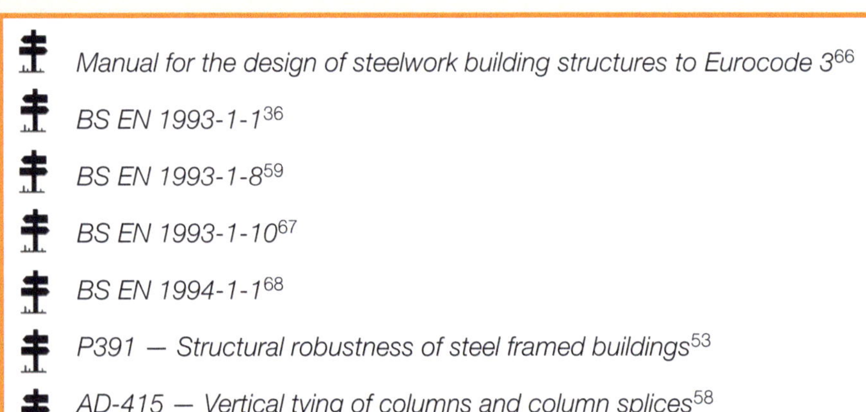

Manual for the design of steelwork building structures to Eurocode 3[66]

BS EN 1993-1-1[36]

BS EN 1993-1-8[59]

BS EN 1993-1-10[67]

BS EN 1994-1-1[68]

P391 — Structural robustness of steel framed buildings[53]

AD-415 — Vertical tying of columns and column splices[58]
This should be read in conjunction with SCI P391

8 Design of light steel frame buildings and modular construction

8.1 Introduction to light steel frame (LSF) buildings

Light steel framing is generally based on C- or Z-sections, cold-rolled from thin steel plate and galvanised.

Light steel frame has many uses within the built environment. It is used in loadbearing frames, as steel studs and as rafters and purlins. It is also used as infill to structural steel or concrete frames, where its function is not loadbearing other than to support or restrain cladding and to resist wind load.

Cold-formed steel sections are common in mezzanines within industrial buildings and are also used in hybrid construction, where loadbearing light steel frames form residential accommodation, supported on a storey of reinforced concrete or steel construction.

In portal frames formed from hot-rolled universal beam or universal column sections, cold-formed sections are used for cladding rails or purlins, which perform a secondary function of providing restraint to the main portal frame.

This chapter follows the requirements of Eurocode 3[36] and BS EN 1991-1-7[4] and the supplementary National Annexes[57,33]. Eurocode 3 contains no specific requirements for robustness, so it is necessary to refer to Annex A in BS EN 1991-1-7 for guidance. Other key references are SCI Publication P402[69] and SCI Technical Information Sheet ED021[70].

8.2 Summary of requirements for each consequence class

If CC1 buildings are designed to Eurocode 3, no additional measures are required.

CC2a buildings should be provided with horizontal ties as defined in A.5.1 of BS EN 1991-1-7.

For CC2b buildings, horizontal ties, as defined in A.5.1 for framed buildings, together with vertical ties, as defined in A.6, should be provided in all supporting columns and walls.

Alternatively, for CC2b buildings, the notional element removal or key element approach can be taken. Horizontal ties will still be required. It is very difficult to envisage how a successful result would be achieved (limited area of collapse) from notional element removal if there is no tying in the structure, or how a structural element could sustain $34kN/m^2$ if it were not tied into the rest of the frame.

8.3 Tying

BS EN 1991-1-7 presents generic robustness rules for all forms of construction and does not provide specific rules for light steel framed buildings. However, the UK National Annex to BS EN 1991-1-7 does present additional guidance for lightweight construction.

Clause NA.3.1 states:

"In the case of lightweight building structures (e.g., those whose primary structure is timber or cold-formed thin gauge steel) the values for minimum horizontal tie forces in Expressions A.1 and A.2 should be taken as 15kN and 7.5kN respectively."

For floor joist ties: $0.8(g_k + \Psi_1 q_k)\, s\, L$ but not less than 5kN/m width perpendicular to the span of the joist
For internal ties: $0.8(g_k + \Psi_1 q_k)\, s\, L$ but not less than 15kN
For peripheral ties: $0.4(g_k + \Psi_1 q_k)\, s\, L$ but not less than 7.5kN

Where:
g_k =   characteristic value of permanent action per unit area of floor or roof (kN/m^2)
L =   length of tie between vertical supports (m)
q_k =   characteristic value of single variable action (imposed floor or roof load) per unit area (kN/m^2)
s =   mean transverse spacing of ties (m)
Ψ_1 =   combination factor for accidental design situations from EN 1990

This is in line with the guidance given in the, now withdrawn, cold-formed light gauge steel code, BS 5950-5[71], and represents a significant reduction from the minimum tie force of 75kN required for other forms of construction.

8.3.1 Horizontal ties
Horizontal ties should be provided in two perpendicular directions.

One set of ties will be formed from the joists themselves, and their end connections should be designed for the required tie force.

The ties orthogonal to the joists can be formed from the head track of the supporting wall below or a Z-section used to support the floor joists, whether situated internally or along the periphery of the building.

Connections are typically formed by screwing or bolting sections together, sometimes with additional angle cleats. In all cases the tie element, any splices and its end connections should be designed for the required tie force.

Horizontal tying members at the periphery of the building should be connected back to the rest of the structure.

8.3.1.1 Additional recommendations for connecting horizontal structure to vertical structure
In a hybrid structure, if the main structural elements are discrete columns (such as hot-rolled steel or a cluster of light steel studs), the SCI provides additional advice. It is recommended that the horizontal ties anchoring the columns nearest to the edge of a floor or roof should be capable of resisting a factored tensile load, acting perpendicular to the edge, equal to the greater of the load for an internal tie, or a minimum of 1% of the factored vertical design load in the column acting at that level.

If the vertical loads are resisted by a distributed assembly of closely-spaced elements, the tying members should be similarly distributed to ensure that the entire assembly is effectively tied and their required capacity taken as 1% of the factored vertical load.

8.3.2 Vertical ties
All splices in primary vertical elements, and connections between vertically adjacent wall panels should be capable of resisting a tensile force equal to the largest design vertical permanent and variable load reaction applied to the column (or wall) from any one storey.

$$T_v = (G_k + \Psi_1 \times Q_k)$$

Wall panels are typically screwed or bolted through the base track to the wall panel below, to satisfy the vertical tying requirement.

8.4 Notional removal of structural elements

If the conditions for the tying method are not met, the structural engineer should check the structure to ensure that disproportionate collapse would not be triggered by the notional removal of vertical loadbearing elements, considered one at a time and across the whole structure.

If the light steel frame is a panel structure, notional removal requires the structure to be checked for the removal of a length of wall 2.25 times the storey height. The load combination to check that the structure does not collapse disproportionately is given in Expression 6.11b of BS EN 1990, simplified to:

$$G_k + \Psi_{1,1} \times Q_{k,1} + \Psi_{2,2} \times Q_{k,2}$$

Refer to section 7.4.2 for further discussion of Expression 6.11b.

8.5 Key element design

If the notional removal of a vertical loadbearing element would risk the collapse of an area greater than the allowable limit of the minimum of 15% of floor area or 100m^2, then that vertical element should be designed as a key element for an accidental load of 34kN/m^2.

The load combinations for accidental actions should be used with the corresponding partial load factors, using Expression 6.11b.

It cannot be assumed that plasterboard will still be intact after an accidental event and therefore the lateral restraint provided by the plasterboard should be ignored.

8.6 Modular buildings

Robustness for modular buildings is provided by the ties between the modules. It can be very difficult to tie a modular building fully due to the lack of access to form connections. Access must be carefully considered at the design stage and access limitations may require innovative connection solutions (Figure 8.1).

Figure 8.1: Access to connections in modular buildings

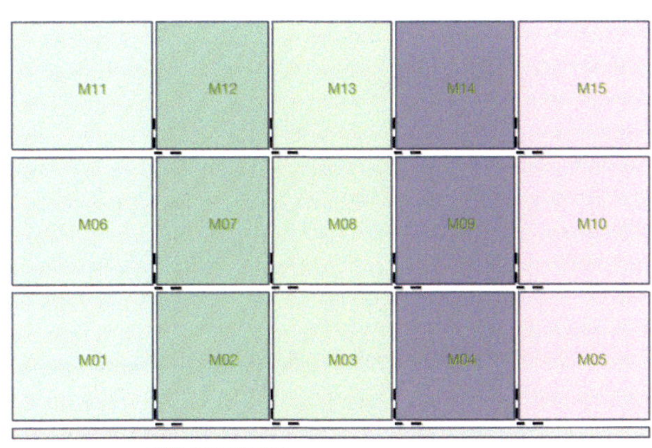

A typical approach would be to provide horizontal ties and check for element removal of panel lengths of 2.25h and full module loss (Figure 8.2). The box-like nature of a modular building, and the close spacing of posts within a panel make notional element removal a feasible approach for checking against disproportionate collapse (Figure 8.3).

Figure 8.2: Modular buildings — modules lost

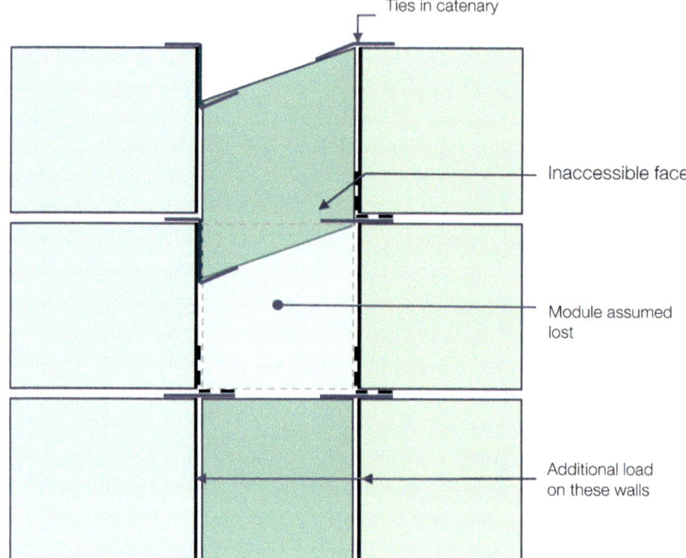

If taking the approach of providing horizontal and vertical ties, the SCI recommends a minimum tying force of half the weight of a module, but not less than 30kN[70].

Figure 8.3: Tying forces in modular construction subject to notional removal of one module

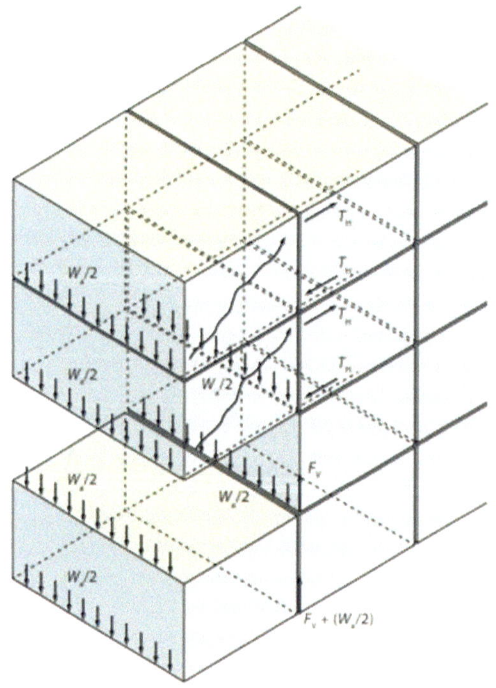

P402 — Light steel framing in residential construction[69]

Robustness of light steel construction[70]

Robustness of light steel frames and modular construction[72]

P348 — Building design using modules[73]

Case study 4

Five-storey, light steel frame building — tying and other considerations

The building is a five-storey residential building (Figure 8.4). It is classed as a CC2b building. The floors and roof are lightweight, consisting of light steel frame (LSF) joists with plasterboard and decking. There are no partition loads on the floor, as every wall is loadbearing LSF.

Figure 8.4: Plan of building

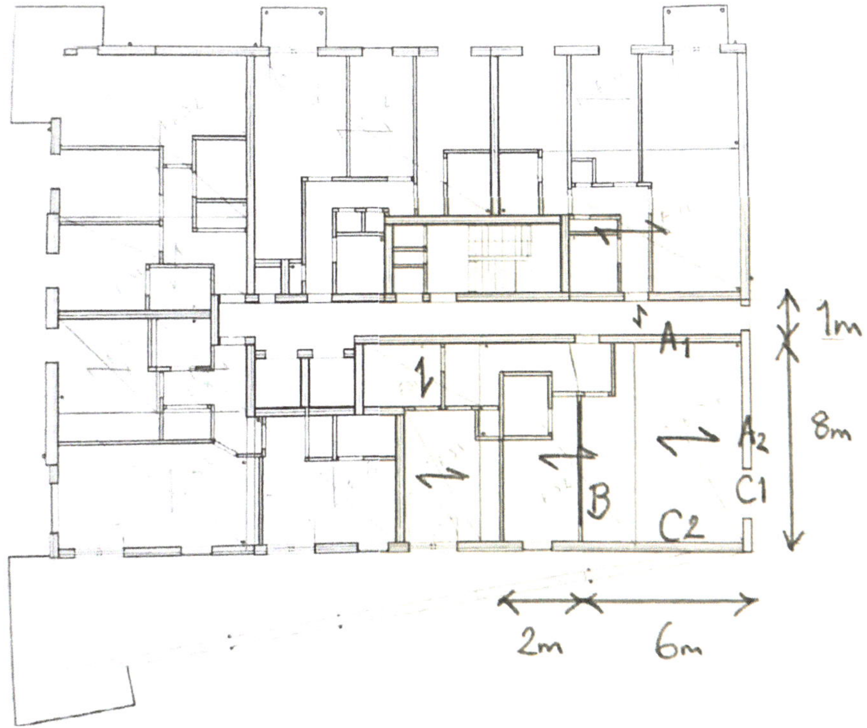

The building is checked to see if the structure is effectively tied by the connections between structural members.

The floor permanent loading is $1kN/m^2$ and floor variable loading is $1.5kN/m^2$.

At roof level the permanent loading is $1kN/m^2$ and variable loading is $0.75kN/m^2$.

This study considers the shaded area in Fig. 8.4. The joists are 250mm deep at 600mm centres.

Horizontal ties
Floor joist ties
Tie force $= 0.8(g_k + \Psi q_k)\, s\, L$ but not less than 5kN/m perpendicular

$$= 0.8 \times (1kN/m^2 + 0.5 \times 1.5kN/m^2) \times 0.6m \times 6m = 5kN \text{ in each joist}$$

This equates to 5kN/0.6m = 8.5kN/m, so is not less than the minimum of 5kN/m.

This can be achieved with a cleat connection at the end of the joist screwed to a Z-section on internal and perimeter walls.

Internal ties

The internal tie is the top of the internal wall, labelled B. The internal tie will be formed by the head track of the wall, with a plate connection where there is a discontinuity in head track.

Tie force = $0.8(g_k + \Psi q_k) \, s \, L$ but not less than 15kN

$$= 0.8 \times (1kN/m^2 + 0.5 \times 1.5kN/m^2) \times (6m + 2m)/2 \times 8m = 45kN$$

Wall B twice turns the corner before it meets the corridor wall and the engineer should consider whether to continue the tie straight through to the corridor, or follow the wall as it changes direction.

A cleat connection at any splice and at the end of the tie will be most suitable.

Peripheral ties

The peripheral ties are at the top of the external walls, labelled C1 and C2.

Tie force, C1 = $0.4(g_k + \Psi q_k) \, s \, L$ but not less than 15kN

$$= 0.4 \times (1kN/m^2 + 0.5 \times 1.5kN/m^2) \times 6m \times 8m = 34kN$$

Tie force, C2 = $0.4(g_k + \Psi q_k) \, s \, L$ but not less than 15kN

$$= 0.4 \times (1kN/m^2 + 0.5 \times 1.5kN/m^2) \times 8m \times 6m = 34kN$$

A cleat connection at any splice will be suitable.

Vertical ties

The cladding is masonry and not supported by the LSF. Therefore, no cladding loads need to be considered in this calculation. For simplicity, the internal plasterboard has also been ignored in these workings.

At A1, the studs and splice details should be capable of supporting a force of:

$g_k + \Psi_1 \times q_k = (1kN/m^2 + 0.5 \times 1.5kN/m^2) \times 0.6m \times 1m/2 = 0.5kN$ (at a spacing of 0.6m)

At A2, the studs and splice details should be capable of supporting a force of:

$g_k + \Psi_1 \times q_k = (1kN/m^2 + 0.5 \times 1.5kN/m^2) \times 0.6m \times 6m/2 = 3kN$ (at a spacing of 0.6m)

This can be achieved with screws at any splices and at the connection of the stud to the top and bottom track of the wall.

Further checks in accordance with SCI guidance

"The horizontal ties anchoring the columns nearest to the edge of a floor or roof should be capable of resisting a factored tensile load, acting perpendicular to the edge, equal to the greater of the load for an internal tie, or a minimum of 1% of the factored vertical design load in the column acting at that level."

First, check the stud at A2:

The building is five storeys and the factored vertical load is:

$(4 \times [1.35 \times 1kN/m^2 + 1.5 \times 1.5kN/m^2] + 1 \times [1.35 \times 1kN/m^2 + 1.5 \times 0.6kN/m^2]) \times 0.6m \times 6m/2 = 31kN$

1% of 31kN = 0.3kN

Check the floor joist ties: 5kN is greater than 0.3kN so this is acceptable.

Check the internal tie, because it connects to the edge of the building: 45kN is greater than 0.3kN so this is acceptable.

9 Design of *in situ* concrete buildings

9.1 Introduction to *in situ* concrete buildings

In situ reinforced concrete is used across all sectors of the building industry, and offers a wide range of structural forms, most commonly in multi-storey concrete buildings of column and flat slab construction, or column and floor beam construction. Resistance to wind load is generally provided by shear walls or cores around lifts and stairs.

In situ concrete is also used in buildings of hybrid construction, frequently with concrete being used for ground-floor columns and a first-floor transfer structure, with a steel frame or masonry structure over.

Although well-detailed *in situ* concrete is inherently robust, and the prescriptive tying rules for concrete are easy to calculate and implement, it is still important to be aware of significant support structures, such as cantilevers, transfer beams and slabs, which may need to be designed as key elements. In most cases, checks for compliance with tying rules will show that tying requirements have already been met through the standard reinforcement provided. Therefore, this chapter mostly discusses best practice in positioning and detailing of tie reinforcement, rather than how to provide for minimum code compliance.

In situ concrete is frequently used for car park construction, and in these cases vehicle impact loads need to be considered, even if the structure is designed with the prescriptive tying rules[64].

This chapter follows the requirements of Eurocode 2[37], the supplementary National Annex[38] and PD 6687-1[65]. The tie forces calculated with the National Annex are different from those in Eurocode 2 and align the forces, with the exception of the vertical tie force and peripheral tie force, with the older (now withdrawn) British Standard, BS 8110-1[39].

Eurocode 2 requires concrete structures to have a suitable tying system if they are not designed to withstand accidental actions. It sets out guidance on how to achieve a suitable tying system which is intended as a minimum requirement, rather than an addition to the reinforcement design based on analysis. The Eurocode requirements for tying do not refer to different consequence classes and, therefore, may be considered to apply to all concrete structures. It is recommended that for Consequence Class 2b buildings, horizontal ties are provided, even when adopting a key element or notional removal approach.

9.2 Notional horizontal load

The Eurocodes set out the requirement for structures to resist a notional horizontal load, represented by a notional inclination, in addition to applied horizontal loads, such as wind loading. The notional inclination leads to all vertical actions having a corresponding horizontal action, with the same load factor and combination factor as the associated vertical load.

Figure 9.1 provides the notional inclination for a range of building heights and column numbers from the IStructE's *Manual for the design of concrete building structures to Eurocode 2*[74].

9.3 Tying

9.3.1 Provision of ties

Four types of ties are specified in BS EN 1992-1-1[37]: peripheral ties, internal ties, column/wall ties and vertical ties (Figure 9.2).

Figure 9.1: Notional inclination of a structure

At the ultimate limit state the horizontal forces to be resisted at any level should be the sum of:

i) The horizontal load due to the vertical load being applied to a structure with a notional inclination. This inclination can be taken from Table 3.1. This notional inclination leads to all vertical actions having a corresponding horizontal action. This horizontal action should have the same load factor and combination factor as the vertical load it is associated with.

ii) The wind load derived from BS EN 1991-1-4 multiplied by the appropriate partial safety factor.

The horizontal forces should be distributed between the strongpoints according to their stiffness and plan location.

Table 3.1 Notional inclination of a structure

Building height (m)	Number of columns stabilised by bracing system			
	1	5	10	≥ 20
≥ 10	1/300	1/390	1/410	1/410
7	1/270	1/340	1/360	1/370
4	1/200	1/260	1/270	1/280

Note

These values are derived from Expression (5.1) of EC2.

Figure 9.2: Horizontal ties for accidental actions

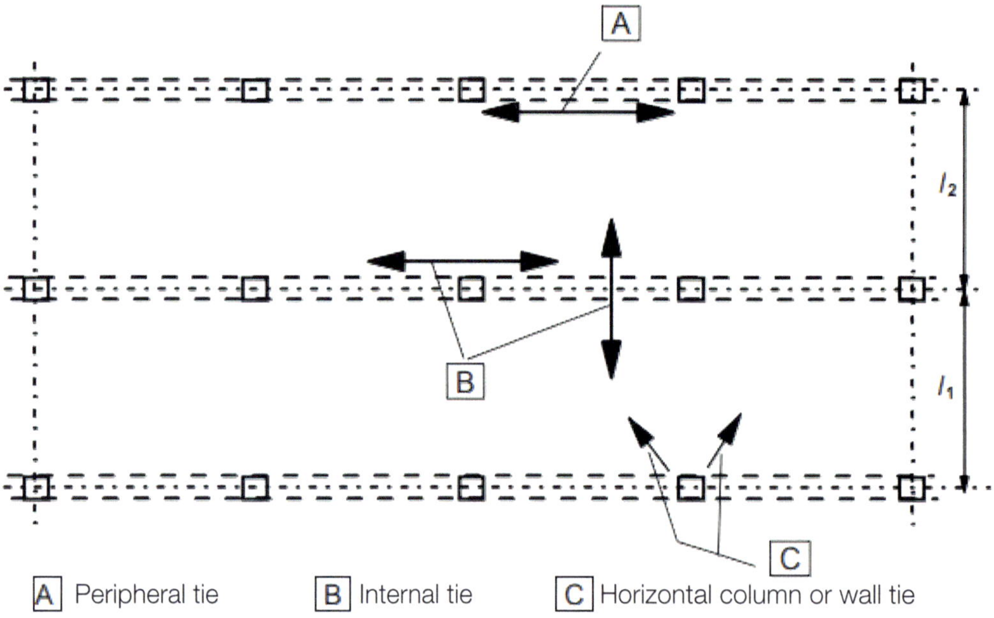

A Peripheral tie | B Internal tie | C Horizontal column or wall tie

Ties are usually assumed to act at their characteristic strength (using a partial factor of $\gamma_s = 1$ for reinforcement and prestressing steel), and no other actions are considered in conjunction with the tie force.

This means that bars provided for other structural effects can be included in the area of reinforcement assumed for the tie — ties are intended as a minimum rather than as additional reinforcement to that required by structural analysis for moment, shear, compression and tension.

Partial factors for materials are set out in Table 9.1 to be used in the calculation of tie capacity.

Table 9.1: Partial factors for materials for ultimate limit states

Design situations	γ_c for concrete	γ_s for reinforcing steel	γ_s for prestressing steel
Persistent and transient	1.5	1.15	1.15
Accidental	1.2	1.0	1.0

For maximum benefit, ties should be formed of higher ductility reinforcement, e.g., class B or C, and horizontal ties should ideally be placed towards the bottom of the section, as tests and experience have shown bottom bars are more effective.

9.3.2 Peripheral ties
The edge of a structure is probably the most vulnerable to damage and has reduced opportunities for developing alternative load paths via two-way spanning.

Provision of a peripheral tie ensures an alternative load system in the edge of the structure. The peripheral tie also provides a zone in which internal ties can be anchored, and ensures that perimeter vertical elements are interconnected with the main tying system. Peripheral ties should also be provided around any large slab openings, such as those for atria.

The UK National Annex defines the tie forces to be resisted, and these are in line with the values given in BS 8110-1 except for the peripheral tie force. The minimum peripheral tie force value $F_{tie,per}$ is 60kN (ultimate) which is easily carried by two H10 bars. The resistant reinforcement should be located within 1.2m from the edge of the structure.

Peripheral tie force is expressed in kN

$$F_{tie,per} = l_i\,q_1 \geq Q_2 \qquad\qquad \text{(Expression 9.15, BS EN 1992-1-1)}$$

$$q_1 = (20 + 4n_o)/l_i \qquad\qquad \text{(Subclause 9.10.2.2 (2) of the UK National Annex)}$$

$$Q_2 = 60\text{kN} \qquad\qquad \text{(Subclause 9.10.2.2 (2) of the UK National Annex)}$$

Where:
n_o = total number of storeys in the structure
l_i = length of the end-span

Therefore:

$$F_{tie,per} = (20 + 4n_o) \geq 60$$

For peripheral ties, the expression for $F_{tie,per}$ comes from the Eurocode which was amended to make Q_2 a minimum rather than a maximum. This means that the value is not consistent with BS 8110 or the values for internal and horizontal ties. In practical terms, the reinforcement required based on the structural analysis will often be sufficient to meet the peripheral tie requirements, so the discrepancy is unlikely to lead to any change in reinforcement provision.

This amendment is repeated in Expression 9.16, which is for internal ties on floors without screeds.

Many other publications are incorrect, making assumptions that there was no difference between BS 8110 and Eurocode 2 peripheral tie expression and, therefore, frequently switching the direction of the '$\geq$'. Engineers need to be cautious.

9.3.3 Internal ties

Internal ties should be provided in two orthogonal directions and should be anchored to peripheral ties at each end, unless they continue as horizontal ties to columns or walls.

The UK National Annex to BS EN 1992-1-1 defines the internal tie forces to be resisted, and these are in line with BS 8110-1. The UK National Annex also gives a maximum spacing for internal transverse ties of $1.5l_r$, where l_r is defined as the greater of the distances between the centres of the columns, frames or walls supporting any adjacent floor spans, in the direction of the tie under consideration. Figure 9.3 illustrates a layout where intermediate ties are required to meet spacing limits.

However, it is generally beneficial to adopt a lower spacing; indeed the requirement for these ties to interact with column ties means that a practical maximum is the column spacing. For *in situ* concrete frames, ties are typically provided by reinforcement distributed across the slab, rather than concentrated on column lines.

Figure 9.3: Column layout where intermediate ties are required to meet spacing limits

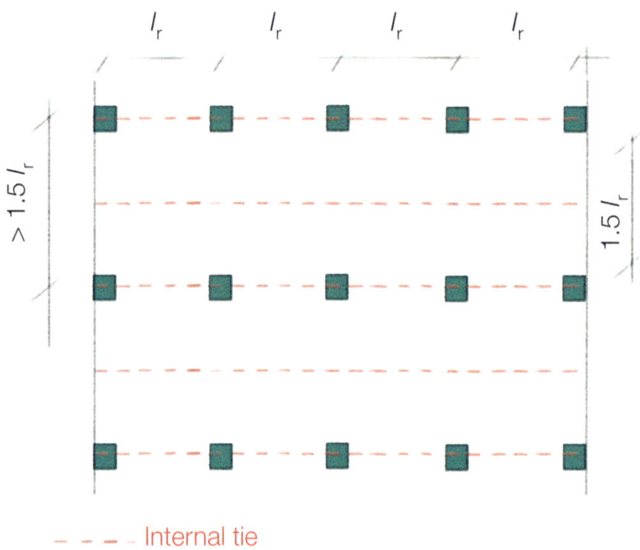

$1.5l_r$ is the maximum spacing between ties
As distance is $>1.5l_r$ intermediate ties are needed

For internal ties, the force is derived in kN/m width (of the slab), whereas for peripheral ties the force is confined to a defined width at the edge of the slab. Internal ties can be spread evenly throughout slabs or may be grouped at/in beams or walls. In walls they should be within 0.5m of the top or bottom of the floor slab.

Internal tie force:

$F_{tie,int}$ is defined in kN/m across the internal slab width and is the greater of $(1/7.5)\,(g_k + q_k)\,(l_r/5)\,F_t$ or F_t

Where:
$(g_k + q_k)$ = the sum of the average permanent and variable floor loads (kN/m^2). Note the variable load q_k is not in this case reduced
l_r = the greater of the distances (m) between the centres of the columns, frames or walls supporting any two adjacent floor spans, in the direction of the tie under consideration

$F_t = (20 + 4n_o) \leq 60$ (where n_o is the total number of storeys in the structure)

9.3.4 Horizontal ties to external walls and columns

Edge columns and walls should be tied to the horizontal structure at each floor and roof level, except in cases where the peripheral tie is located within the wall. For corner columns, the tie force should be provided in two directions approximately at right angles to each other.

The UK National Annex to BS EN 1992-1-1 defines the horizontal tie forces to be resisted and these are in line with BS 8110-1.

$F_{tie,fac}$ is in kN per m length of wall. $F_{tie,col}$ is in kN per column.

Tie force:

$F_{tie,fac} = F_{tie,col}$ = greater of $2F_t$ ($\leq (L_s/2.5) F_t$) and 3% of the total design ultimate vertical load carried by the column or wall at that level.

Or expressed more simply as:

$F_{tie,fac} = F_{tie,col}$ = Maximum (Minimum ($2F_t$; $l_s F_t/2.5$); $0.03 N_{Ed}$)

Where:
l_s =   floor-to-ceiling height in m

$F_t = (20 + 4n_o) \leq 60$ (where n_o is the total number of storeys in the structure)

9.3.5 Vertical ties

Vertical ties should be provided to every column and to each wall carrying vertical load, from the lowest to the highest level.

The tie force to be resisted is equal to the design load carried on any floor level by the element calculated with the accidental load factors (Section 5.8 and Tables NA.A1.1 and NA.A1.3 National Annex of Eurocode 0[44]).

The loading in the accidental design situation may be calculated from the following expression:

$$E_d = G_k + A_d + \Psi_{1,1}Q_{k,1} + \Sigma \Psi_{2,i}Q_{k,i}$$

Where:
E_d =   design value of the effect of actions
G_k =   characteristic value of a permanent action
A_d =   design value of an accidental action
Ψ_1 =   frequent value of a variable action
Q_k =   characteristic value of a variable action
Ψ_2 =   quasi-permanent value of a variable action

When using this expression, the accidental load, A_d, would be taken as 0 because the Eurocode recommends the use of the expression after the accident has taken place.

Although Eurocode 2 only requires vertical ties for buildings of five storeys and above, in the UK Approved Document A overrides this, and vertical ties should be provided for all buildings in CC2b and above.

Where the vertical element is supported at its lowest level by anything other than a foundation, the overall robustness of the support should be considered, and most likely the supporting element will be designed as a key element, or alternatively analysis may be carried out to demonstrate that removal of the supporting element does not lead to disproportionate collapse.

The calculation of vertical tie force is different between BS 8110 and Eurocode 2 in that different load combinations and load factors are used.

Also different is that BS 8110 expected the tie force to be the greatest force calculated from any one storey and applied to the whole column, whereas Eurocode 2 calculates the force at a storey, and applies that tie force to the column below only.

This difference means that the vertical tie forces in concrete and steel designed to the Eurocodes are not aligned. In fact, steel and concrete have moved in opposite directions with regard to whether the greatest force from any storey is used, since the transition from British Standards to Eurocodes.

BS 8110: *"Each column and wall should be provided with ties continuously from lowest to highest level. Ties should be capable of resisting design force equal to maximum ultimate dead and imposed load received by wall and column from any one storey."*

Eurocode 2: *"Continuous ties should be provided from lowest to highest level capable of carrying load in the accidental load situation from floor above column/wall accidentally lost."*

9.3.6 Continuity of ties and good detailing practice

Ties should be effectively continuous. Where bars forming ties are lapped, the detailing should be such that failure is always in the bar and not in the lap, ensuring a ductile performance. In practice, this means that laps should always be designed for the full bar capacity, even if the required tie force is lower.

Additionally, where bars forming the tie system are lapped but not adjacent, the lap length should be increased and the need for links considered, as described in BS EN 1992-1-1, Clause 8.7. This is to ensure that an effective strut and tie system can be formed between the bars in the lap zone.

For tying systems to work, different types of ties must interact, and the need for vertical ties, internal ties and peripheral ties to be linked has already been discussed. It is essential for vertical and horizontal ties to interact if catenary action is to be developed.

There is an implicit assumption of ductile structural performance for the nominal tying recommendations to work in practice, not least since catenary action presumes significant axial and rotational ductility. In routine design, ductility demands are not calculated explicitly; they are catered for by assuring members are under-reinforced and by application of detailing rules. In cases where ductile energy-absorbing capacity is required for extreme robustness (as in resistance to blast or earthquakes), advanced rules are available.

BS 8110-1 stated that connection of horizontal and vertical elements could:

"Generally be achieved by ensuring a minimum of two bars in each direction pass directly through the column".

BS EN 1992-1-1, Clause 9.4.1(3) has a similar requirement, although this is specifically for flat slabs. It is therefore recommended that some horizontal ties in each direction pass through the column, and that they are placed in the bottom of the slab or beam at the column location.

It is also important to note that if the vertical tie is to share load up the building, the connection to the higher floors must be capable of taking reverse shear. If a floor is hung from the column above, the column is pulling down on the floor above, i.e., the shear is reversed from the standard situation, where the column would be pushing up on the floor. It would be hard to determine what shear force to design for, as load will be distributed through the remaining structure. However, for *in situ* concrete construction, if robustness requirements are met through tying, this does not require further consideration as connection details inherently provide shear resistance both up and down. For precast and hybrid type connections further consideration needs to be given on how the connection would perform if shear were to be reversed.

A minimum percentage of reinforcement in columns assures a minimal tensile capacity for columns, even though columns are designed as compression members. Minimum link requirements in columns are defined, to give a certain level of ductility at the column/floor connection.

Similarly, design codes require minimum amounts of reinforcement at supports, even where no moment has been assumed. These requirements, driven by the need to control cracking, also provide alternative (or enhanced) load paths.

Such additional reinforcement provides protective strength against the possibility of moment reversal, which can be a feature in accidental loading.

It can be seen that there are a number of detailing requirements in the design codes which add to the inherent robustness of concrete buildings. It is important that where such minimum reinforcement requirements are not met, e.g., due to new construction techniques or systems, the effect on the overall structural robustness should be reconsidered.

Case study 5

The Murrah Building

The collapse of the Murrah Building in Oklahoma City[75] was triggered by a terrorist attack. However, the event was disproportionate in that only one column was knocked out by the attack, but that triggered a collapse. In robustness terms, the collapse following that single column removal was consequent on having no continuity rebar on the bottom of the beam over the column support (Figure 9.4). This highlights the importance of ensuring that the reinforcement detailing is correct.

Figure 9.4: Blast damage and collapse of the Murrah Building, Oklahoma City

9.3.7 Additional considerations for post-tensioned concrete

The principles for post-tensioned construction are identical to those adopted for reinforced concrete. In bonded prestressed concrete, continuous tendons provide an excellent tie, since there are no (or at least fewer) laps and therefore they can be used for the tying system.

For bonded construction, the main challenge is to ensure sufficient interaction between the horizontal and vertical ties. At columns, tendons are located in the top of the slab and may pull out of the shear cone (Figure 9.5), so it is recommended that additional bottom steel is provided through the column, to lap onto the tendons in the lower half of the slab.

At mid-span, the tendons are in the bottom of the slab and provide support to the structure, so no additional reinforcement is required. Research has also shown that a significant improvement in post-failure capacity occurs when tendons pass directly through the column, so it is recommended that ducts pass through the column or are placed as near as possible (either side of the column) (Fig. 9.5)[76].

Figure 9.5: Catenary action of tendons at column head

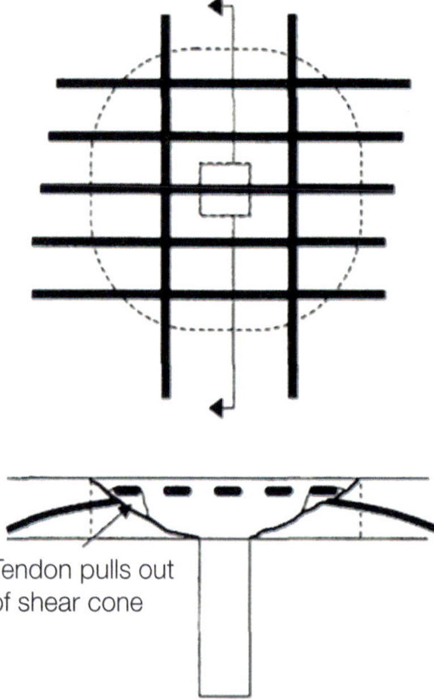

For unbonded tendons, a failure of tendons in one bay may lead to failure in adjacent bays. For this reason, it is not appropriate to consider unbonded tendons as part of the tying system, so tying should be provided wholly with normal reinforcement.

9.4 Notional element removal

An alternative to following the prescriptive rules for tying is to consider notional removal of an element and to investigate the subsequent collapse area following the principles detailed in Chapter 5. There are no special requirements for *in situ* concrete structures, except that partial factors for materials can be reduced and the accidental load case used (Expression 6.11b and Tables NA.A1.1, and NA.1.3 of BS EN 1990).

In BS EN 1992-1-1, for the accidental load case, the partial factors are discussed in Section 9.3.1 and are as follows:

γ_c for concrete = 1.2
γ_s for steel = 1.0

As for all materials, the amount of collapse allowed is limited to 100m² or 15% of the floor area and two adjacent storeys.

9.5 Key element design

There are no special requirements for concrete key element design, and the approaches described in Section 5.7 should be adopted.

Case study 6

Three-storey *in situ* concrete building, provided with horizontal ties

Three-storey residential building (Figure 9.6)

Figure 9.6: Section

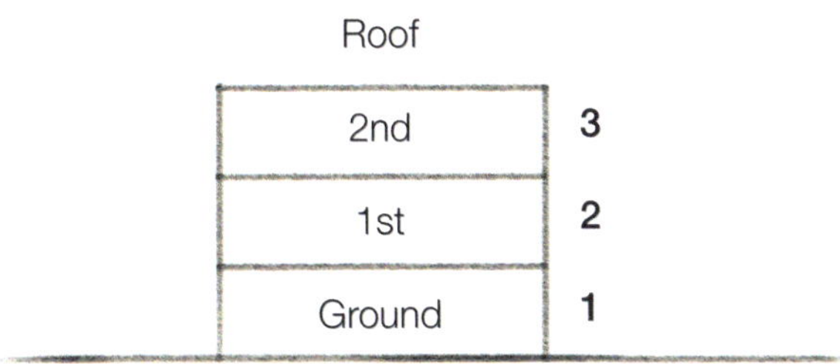

Structure
Flat slab, column grid 7.5 x 7.5m, 3.5m floor-to-floor, 250mm thick with B16 @ 200mm centres each direction, in the top and bottom of the slab.

$F_{tie,cap}$ per bar = $(16/2)^{^2} \times \pi \times 500/1.0 \times 10^{^-3}$ = 100.5kN

Floor loading
Permanent actions, g_k = 0.25m x 25kN/m^2 + 1.0kN/m^2 = 7.25kN/m^2 (all floors)

Variable actions, q_k = 3.5kN/m^2 (residential floors)
Variable actions, q_k = 1.5kN/m^2 (accessible roof)

Ultimate limit state action = 1.35 x 7.25kN/m^2 + 3.5 x 1.5kN/m^2 = 15.04kN/m^2 (floors)
Ultimate limit state action = 1.35 x 7.25kN/m^2 + 1.5 x 1.5kN/m^2 = 12.04kN/m^2 (accessible roof)

Building classification
A residential building not exceeding four storeys is classified as CC2a.

Robustness requirements
CC2a buildings require horizontal ties. Three types of horizontal ties are required: peripheral ties, internal ties and horizontal column/wall ties.

Peripheral tie force, for residential floors

$F_{tie,per}$ = $(20 + 4n_o) \geq$ 60kN

= 20 + 4 x 3 = 32kN

$\Rightarrow F_{tie,per}$ = **60kN**

Where n_o is the total number of storeys in the structure

1 x B16 is sufficient to resist the tie force — therefore the existing slab reinforcement meets tie requirements.

Internal tie force

$F_{tie,int}$ is defined in kN/m across the internal slab width and is the greater of:

$(1/7.5)\,(g_k + q_k)\,(l_r/5)\,F_t = (1/7.5) \times (7.25\text{kN/m}^2 + 3.5\text{kN/m}^2) \times (7.5\text{m}/5) \times 32 = 69\text{kN/m}$

Or $F_t = (20 + 4n_o) = 32\ (\leq 60)$

$\Rightarrow F_{tie,int} = \mathbf{69kN/m}$

Where:
$(g_k + q_k)$ is the sum of the average permanent and variable floor loads (kN/m^2)
  Note the variable load, q_k, is not in this case reduced (Section 5.8)
l_r is the greater of the distances (m) between the centres of the columns, frames or walls supporting any two adjacent floor spans in the direction of the tie under consideration

$F_t = (20 + 4n_o) \leq 60$ (where n_o is the total number of storeys in the structure)

B16 bars at 200mm centres are sufficient to resist tie force — therefore the existing slab reinforcement meets tie requirements.

Horizontal tie force (level considered — first floor)

$F_{tie,fac} = F_{tie,col} = \text{Maximum (Minimum } (2F_t;\ l_s\,F_t/2.5);\ 0.03\,N_{Ed})$

$F_t = (20 + 4n_o) \leq 60$ (where n_o is the total number of storeys in the structure)

Where:
$F_{tie,fac}$ is in kN per m length of wall
$F_{tie,col}$ is in kN per column
l_s is the floor-to-ceiling height in m

$F_t = (20 + 4n_o) \leq 60$ — as above — 32kN. Therefore $2F_t = 64$kN

$(L_s/2.5)F_t = 3.5\text{m}/2.5 \times 32\text{kN} = 45\text{kN}$

3% of the total design ultimate vertical load carried by the column or wall at that level (in this case the first-floor level)

Edge column: vertical load = $7.5\text{m} \times 7.5\text{m}/2 \times (2 \times 15.04\text{kN/m}^2 + 1 \times 12.04\text{kN/m}^2) = 1{,}185\text{kN}$

3% = 36kN

Corner column: vertical load = $7.5\text{m}/2 \times 7.5\text{m}/2 \times (2 \times 15.04\text{kN/m}^2 + 1 \times 12.04\text{kN/m}^2) = 592\text{kN}$

3% = 18kN

$F_{tie,fac} = \text{Maximum (Minimum } (64\text{kN};\ 45\text{kN});\ 36\text{kN or } 18\text{kN})$

Therefore, $F_{tie,fac} = \mathbf{45kN/m}$, $F_{tie,col} = \mathbf{45kN}$ (Cladding loads have been ignored for simplicity)

1 x B16 is sufficient to resist tie force — therefore existing slab reinforcement meets tie requirements.

Conclusion
No change is required to the design to meet tying requirements for the structure.

Case study 7

Ten-storey *in situ* concrete building — tying

Ten-storey office building (Figure 9.7)

Figure 9.7: Section

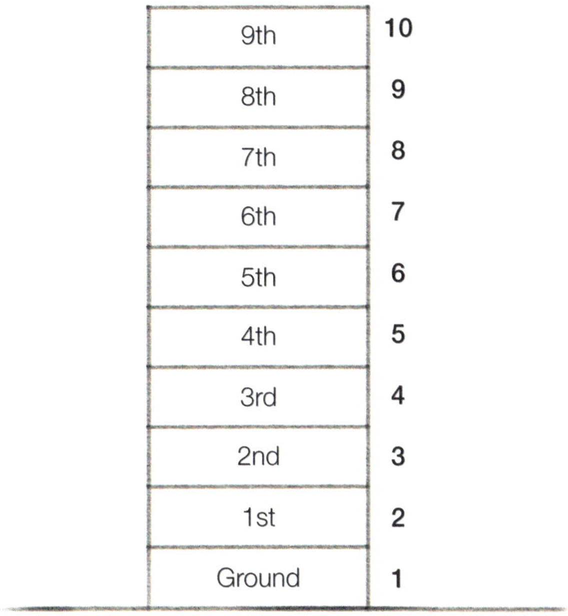

Floor	Storey
9th	10
8th	9
7th	8
6th	7
5th	6
4th	5
3rd	4
2nd	3
1st	2
Ground	1

Structure

Beam and slab, column grid 6m x 9m, 3.5m floor-to-floor, 250mm thick slab (spanning 6m) with B16 @ 200mm centres bottom, 600mm wide x 600mm deep beam with 8 x B32 bars, maximum column size 650 x 650mm, 8 x B32.

$F_{tie,cap}$ per B16 bar = $(16/2)^2$ x π x 500/1.0 x 10^{-3} = 100.5kN
$F_{tie,cap}$ per B32 bar = $(32/2)^2$ x π x 500/1.0 x 10^{-3} = 402kN

Floor loading

Permanent actions, g_k = (0.25m x 25kN/m^2) + 1kN/m^2 + (0.6m x 0.35m x 9m/(6m x 9m) x 25kN/m^3)
   = 8.125kN/m^2 (average value accounting for self-weight of beams)

Variable actions, q_k = 4kN/m^2 + 1kN/m^2 = 5kN/m^2 (floors)
Variable actions, q_k = 1.5kN/m^2 (accessible roof)

Ultimate limit state action = 1.35 x 8.125kN/m^2 + 1.5 x 5kN/m^2 = 18.5kN/m^2 (office floors)
Ultimate limit state action = 1.35 x 8.125kN/m^2 + 1.5 x 1.5kN/m^2 = 13.2kN/m^2 (accessible roof)
     (based on Expression 6.10 in BS EN 1990)

Ultimate limit state action on internal beams: 6m x 18.47kN/m^2 = 110.82kN/m

Building classification

An office building between four and 15 storeys is classified as CC2b.

Robustness requirements

A tying approach will be adopted — CC2b buildings require horizontal and vertical ties. Four types of ties are required: peripheral ties, internal ties, horizontal column/wall ties and vertical ties.

Peripheral tie force

$$F_{tie,per} = (20 + 4n_o) \geq 60\text{kN}$$

$$= 20 + 4 \times 10 = 60\text{kN} \Rightarrow \mathbf{60kN}$$

Where n_o is the total number of storeys in the structure

1 x B16 (or B32) is sufficient to resist the tie force — therefore the existing slab and beam reinforcement meets tie requirements.

Internal tie force

$F_{tie,int}$ is defined in kN/m across the internal slab width and is the greater of:

$$(1/7.5)\,(g_k + q_k)\,(l_r/5)\,F_t$$

For 6m span: $F_t = (1/7.5) \times (8.125\text{kN/m}^2 + 5\text{kN/m}^2) \times (6\text{m}/5) \times 60 = 126\text{kN/m}$

For 9m span: $F_t = (1/7.5) \times (8.125\text{kN/m}^2 + 5\text{kN/m}^2) \times (9\text{m}/5) \times 60 = 189\text{kN/m}$

Or $F_t = 60\text{kN/m}$

$\Rightarrow$ **126kN/m** for 6m span and **189kN/m** for 9m span

Where:
$(g_k + q_k)$ is the sum of the average permanent and variable floor loads (kN/m^2)
Note the variable load, q_k, is not in this case reduced (Section 5.8)
l_r is the greater of the distances (m) between the centres of the columns, frames or walls supporting any two adjacent floor spans in the direction of the tie under consideration

$F_t = (20 + 4n_o) \leq 60$ (where n_o is the total number of storeys in the structure)

2 x B16 or 1 x B32 are sufficient to resist tie force — therefore existing slab and beam reinforcement meets tie requirements.

Horizontal tie force (level considered — first floor)

$$F_{tie,fac} = F_{tie,col} = \text{Maximum (Minimum } (2F_t;\ l_s\,F_t/2.5);\ 0.03\,N_{Ed})$$

$F_t = (20 + 4n_o) \leq 60$ (where n_o is the total number of storeys in the structure)

Where:
$F_{tie,fac}$ is in kN per m length of wall
$F_{tie,col}$ is in kN per column
l_s is the floor-to-ceiling height in m

$F_t = (20 + 4n_o) \leq 60$ — as above — 60kN

$2F_t = 120\text{kN}$

$l_s\,F_t/2.5 = 3.5\text{m} \times 60\text{kN}/2.5 = 84\text{kN}$

3% of the total design ultimate vertical load carried by the column or wall at that level (in this case the first-floor level)

Edge column: vertical load = 9m x 6m/2 x (9 x 18.47kN/m^2 + 1 x 13.22kN/m^2) = 4,845kN

3% = 145kN

Corner column: vertical load = 9m/2 x 6m/2 x (9 x 18.47kN/m^2 + 1 x 13.22kN/m^2) = 2,423kN

3% = 73kN

For the edge column: $F_{tie,fac}$ = $F_{tie,col}$ = Maximum (Minimum (120kN; 84kN); 145kN)

$F_{tie,fac}$ = **145kN/m**
$F_{tie,col}$ = **145kN**

For the corner column: $F_{tie,fac}$ = $F_{tie,col}$ = Maximum (Minimum (120kN; 84kN); 73kN)

$F_{tie,fac}$ = **84kN/m**
$F_{tie,col}$ = **84kN**

2 x B16s or 1 x B32 are sufficient to resist tie force — therefore existing slab and beam reinforcement meets tie requirements.

Vertical tie force

g_k = 0.25m x 25kN/m^3 + 1kN/m^2 + 0.6m x 0.35m x 9m/(6m x 9m) x 25kN/m^3 = 8.125kN/m^2

A_d = 0

g_k = 4kN/m^2 + 1kN/m^2 = 5kN/m^2

Ψ_1 = 0.5 (Table NA.A1.1 from the UK National Annex to EN 1990)

E_d = 8.125kN/m^2 + 0 + 0.5 x 5kN/m^2 = 10.63kN/m^2

Tie force = 10.63kN/m^2 x 9m x 6m = **574kN** $\Rightarrow$ 2 x B32 bars

Therefore, existing column reinforcement meets tie requirements.

Conclusion
No change is required to the design to meet tying requirements for the structure.

Case study 8

Ten-storey *in situ* concrete building — key element design

Ten-storey office building

Structure
Beam and slab, column grid 6m x 9m, 3.5m floor-to-floor, 250mm thick slab (spanning 6m) with B16 @ 200mm centres bottom, 600mm wide x 600mm deep beam with 8 x B32 bars, maximum column size 650 x 650mm, 8 x B32.

$F_{tie,cap}$ per B16 bar = $(16/2)^{2}$ x π x 500/1.0 x 10^{-3} = 100.5kN
$F_{tie,cap}$ per B32 bar = $(32/2)^{2}$ x π x 500/1.0 x 10^{-3} = 402kN

Floor loading
Permanent actions, g_k = (0.25m x 25kN/m^2) + 1kN/m^2 + (0.6m x 0.35m x 9m/(6m x 9m) x 25kN/m^3)
                           = 8.125 kN/m^2

Variable actions, q_k = 4kN/m^2 + 1kN/m^2 = 5kN/m^2 (floors)
Variable actions, q_k = 1.5kN/m^2 (accessible roof)

Accidental   = 34kN/m^2

Building classification
An office building between four and 15 storeys is classified as CC2b.

Robustness requirements
A key element approach will be adopted.

Consider loading in one direction at a time (however, column is square and symmetrical).

Horizontal load, A_d = 34kN/m^2 x 0.65m = 22.1kN/m (based on column width)

Vertical load   = 8.125kN/m^2 + 0.5 x 5kN/m^2 = 10.63kN/m^2 (floors)
Vertical load   = 8.125kN/m^2 + 0 x 1.5kN/m^2 = 8.125kN/m^2 (accessible roof)

Therefore:
N_{Ed} = 10.63kN/m^2 x 9 x 9m x 6m + 8.125kN/m^2 x 9m x 6m= **5,605kN**
M_{Ed} = 22.1kN/m x 3.5m^2/8 = **33.8kNm** (as the column is symmetrical, only one direction needs to be checked)

Ψ_1 is 0.5 for office load and 0 for roofs (Section 5.8). Load factor applied to accidental action is 1.

The column is checked for this axial force and moment.

10 Design of precast concrete buildings

10.1 Introduction to precast concrete buildings

Structures formed of loadbearing precast panels will be forever associated with the Ronan Point collapse, but detailed correctly, precast structures are no less robust than their *in situ* counterparts. However, precast and hybrid constructions lack some of the continuity inherent in *in situ* construction, so the provision of robustness requires more explicit consideration.

This chapter focuses on precast concrete-framed buildings, with additional guidance on discrete structural precast elements such as prestressed hollowcore planks, stair units, balconies and facade panels.

The codified rules regarding robustness of precast framed structures are generally identical to those for *in situ* concrete structures outlined in Chapter 9. This chapter therefore provides explicit guidance, details, and case studies covering specific challenges and differences associated with precast structures, such as connections. Additional guidance is also given regarding anchorage of precast floor and stair units to the supporting structure. It should be noted that fib Bulletin 63[77] sets out analysis procedures for determining more realistic tie forces, which can lead to higher forces than using the codified approach, especially for precast floorplates with no structural topping. Reference to that document should be made in such cases, as the approach is not covered here.

Precast frames typically consist of precast columns, beams, slabs (e.g., solid, hollowcore, lattice plank) and walls (e.g., solid, twinwall, sandwich panel), although some framing systems do not require beams (e.g., flat slab or crosswall). Hybrid systems are also common (e.g., precast cores and columns supporting steel beams). Stability is typically provided through shear walls or cores. Lattice plank and twinwall systems generally allow for fully *in situ* connections and stitches, with design and detailing covered in Chapter 9.

10.2 Design responsibility

BS 8110-1[39] stated:

"The engineer responsible for the overall stability of the structure should ensure the compatibility of the design and details of parts and components, even where some or all of the design and details of those parts and components are not made by this engineer."

This should apply equally to disproportionate collapse considerations, and more so to precast frames, especially if they contain precast elements designed by others (i.e., the precast supplier or manufacturer). The Eurocodes are not as specific, but it is nevertheless good practice for a single party to take overall responsibility for robustness.

10.3 Robustness requirements

The requirements for robustness and preventing disproportionate collapse of any structure, including precast concrete-framed buildings are outlined in Chapter 5, covering notional horizontal load, horizontal and vertical tie forces, notional element removal and key element design, generally referring to BS EN 1991-1-7[4].

The primary requirement for Consequence Class 2b structures is the provision of horizontal ties in two directions, with vertical ties usually also provided, unless the notional element removal or key element design approach is required.

As with *in situ* structures, the full capacity of all reinforcement/connecting parts can be assumed for the tie, regardless of its contribution to structural capacity in-service, so it may be unnecessary to add additional steel in some cases.

Horizontal tying between adjacent structures (e.g., precast wall panels) is not required for robustness, but usually specified for normal in-service design.

The tying requirements for a Consequence Class 2a building are similar, except the vertical ties can be omitted.

10.4 Tying

10.4.1 Horizontal ties

Horizontal tying can generally be accomplished by using the members themselves, with appropriate end connections. Beams designed to carry floor or roof loading will typically be suitable as continuous ties. For example, a series of beams in-line will act as an internal tie, provided end connections to intermediate elements (beams or columns) have the required strength and suitable detailing.

Alternatively, ties may be added as steel members or as steel reinforcement or helical prestressing strand (laid taut but not prestressed), adequately lapped and embedded in *in situ* concrete strips set between precast units, or set within recesses purposely left in the ends of the precast units. Ties within narrow joints (i.e., between precast planks) should be mechanically coupled rather than lapped, and do not constitute anchorage of the floor unit to the beam. This is because floor joints can be opened up due to the deflection induced in an accidental loading situation.

Adequate space must be provided around such bars or strand to ensure good compaction and bond. The minimum spacing is usually at least 50mm. 10mm aggregate is often used to minimise the size of the concrete infill. To maximise the benefit, ties should be formed of higher ductility reinforcement, e.g., class B or C.

Spacing of bars/strand $\geq (\phi + 2H_{agg} + 10mm)$

Where:
ϕ =   bar diameter
H_{agg} =   maximum aggregate size

Beam and column frame solutions require careful detailing to manage the effects of torsion and excessive movements, which can be induced by beam deflection or rotation, and by differential temperature gradients on exposed slabs. Such movements can cause spalling and loss of bearing area, emphasising the need to consider robustness in the detailing.

Where precast floor members form the horizontal ties, they will usually be tied to each other over their support in the direction of their span. When a reinforced structural topping is provided, the mesh provides effective anchorage, but will only contribute to catenary action if it is mechanically connected to the planks, ideally with vertical stirrups anchored in filled cores. Preferably, the topping can be assumed not to act as a tie and anchored U-bars should project from ends of solid planks. For hollowcore, reinforcement should be provided within plank end pockets with bars then grouted up to link units together. Thicker toppings, for example, on top of lattice planks, can contain the ties, as the lattice girders effectively anchor the precast to the topping. Open cores at ends of hollowcore planks should not be more than 600mm long, and adjacent cores should not be opened.

Loops should preferably be in the form of securely anchored hairpins placed in the middle of the floor depth, to allow maximum efficiency and deformability. The anchorage of ties used in catenaries should be demonstrated (to the relevant code) and not simply assumed.

There is no minimum requirement for tying planks to any edge beams/walls that they span parallel to. However, it is best practice to ensure regular ties are achieved (either via the topping or by breaking out cores to form pockets), and this would likely be required for normal in-service design.

10.4.2 Anchorage of precast members

Precast floor, stair or roof members could be dislodged if their supports move due to accidental actions such as an explosion. To prevent this, BS 8110-1, Clause 5.1.8.3 stated that where such elements are not used to provide the horizontal ties, they must still be anchored to the part of the structure containing the ties, with such anchorage being capable of carrying the dead weight of the member.

Eurocode 2 does not cover this but PD 6687-1[65] advises engineers to use the BS 8110-1 requirement for all buildings. This requirement is to prevent either end of a spanning element from becoming free and falling from its support, rather than ensuring that the element can hang in the event that one end does fall.

PD 6687-1 does not require variable actions to be considered when checking robustness, but it is recommended that the dead weight includes permanent finishes, as well as the dead weight imposed on it by supported members (with a partial load factor = 1.0). Where such elements are used in other forms of construction (e.g., steel, timber or masonry) the relevant Eurocode should be followed.

In stair cores, half-landings surrounded by structural walls on three sides can be anchored using proprietary support inserts, which each only transfer shear loads (vertical and horizontal) by locating two in the end wall and one in each side wall[78].

A similar approach is recommended for other discrete precast items that do not form part of the structural system, such as single-skin precast facade panels and balconies, i.e., to ensure that the sum of horizontal restraint capacity back to the structure exceeds the weight of the unit (including finishes). Additionally, it is recommended to confirm that for facade panels if one support point or restraint is removed, the panel would not fall from the building. This may be achieved by ensuring that in the accidental case a facade panel can be supported on the unit below (even if supported floor-by-floor in-service), or by designing remaining support/restraints for increased loading accordingly. Where a panel is base-stacked to the ground with aligned vertical joints (i.e., not staggered), steel plates may be placed in the horizontal joints bridging the vertical joint (and anchored each side with vertical dowel reinforcement) to support panels over a removed panel in the accidental case.

10.4.3 Vertical ties

Vertical tying is the preferred approach for Consequence Class 2b structures, which provides a support system in an accidental event via the suspension of any newly unsupported elements back to an intact structure above the damaged area. This in turn acts through catenary (Figure 10.1) or cantilever action (e.g., in the case of corner column failure where horizontal tie reinforcement in the top of beams can take up the cantilever's tensile stresses).

Figure 10.1: Catenary action in precast floor beams (with composite action)

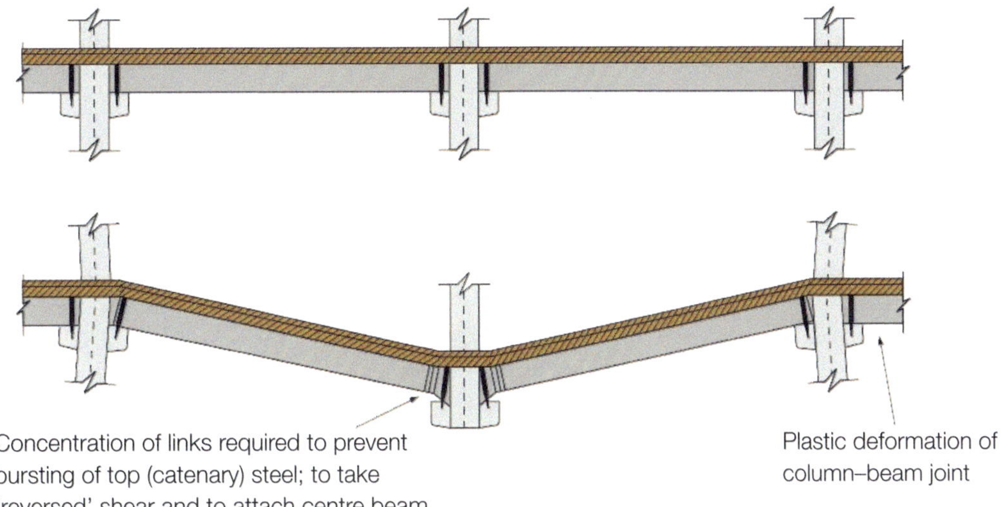

To enable catenary action, longitudinal top reinforcement must be prevented from bursting upwards at the point of maximum sag. To avoid this, links are required as shown in Fig. 10.1, designed for the vertical tie force. The design objective is that stitching the various components together will make the frame behave more like an *in situ* frame in an accidental event.

Under similar loss of support scenarios, prestressed beams spanning onto corbels would likely keep their original shape without significant cracking, and would simply separate from any corbel above the failed column unless anchored down. Rotations and deformations in the column joint zone could be large, so the assumptions on survival are only valid if the anchorage details are sufficiently ductile.

In the unlikely event that vertical tying (Approach 1) is not practical, or where it would not prevent collapse of major parts of the structure in the event of a failure, notional element removal (Approach 2) or key element design (Approach 3) is required. Note that it is best practice still to have vertical ties even if Approaches 2 or 3 are necessary.

10.5 Notional element removal

The notional element removal approach is an appropriate one in precast assemblies made up of discrete elements. This approach will require increased attention to detail, but many panel structures are able to function as deep beams and so inherently provide good spanning capability.

Like frame structures, loadbearing wall structures can also provide alternative load paths. The survival capability of such structures is often high because of the large cantilevering and bridging capability of wall panels, effectively acting as deep beams. The following support systems can operate (Figure 10.2):

- Suspension of the elements from the intact upper structure above the damaged area. This is facilitated by vertical ties from foundation to roof level in all walls
- Cantilever action of the surrounding structure. For example, in the case of corner wall panel failure, the horizontal tie reinforcement on top of the wall panel will take up the tensile stresses of the cantilever. To be effective, the tie reinforcement must be connected into the wall panel, e.g., inside hairpins projecting above the unit tops
- Bridging of the damaged area by the intact wall panels above

Figure 10.2: Alternative means of protection against progressive collapse in wall frame structures

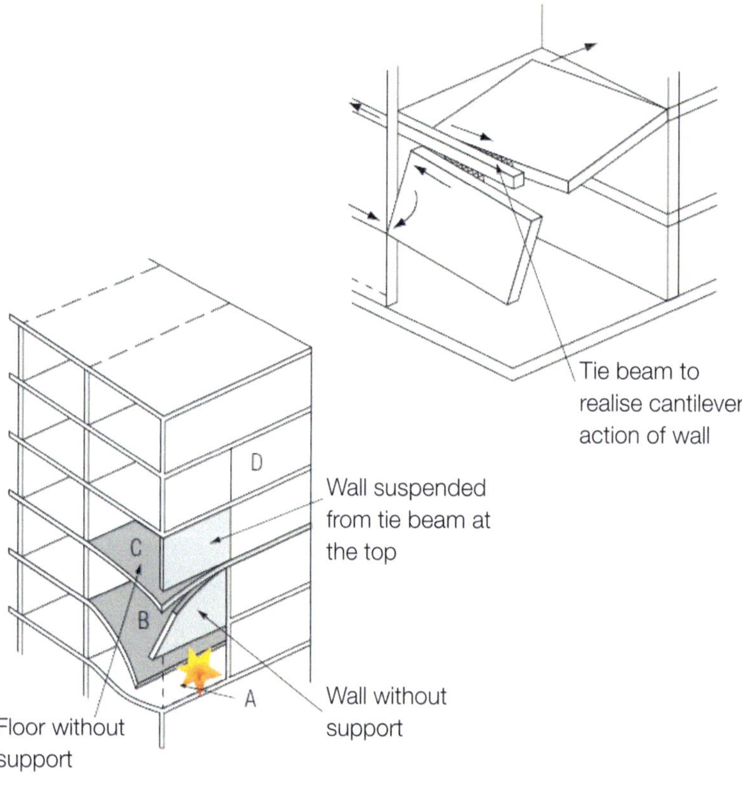

10.6 Typical details

Some proprietary product manufacturers may provide typical details for their own products. Figures 10.3–10.14[3,79] show generic details applicable to common forms of precast construction. The principles are compliant with both Eurocode 2 and BS 8110.

Figure 10.3: Internal floor ties within bonded concrete topping

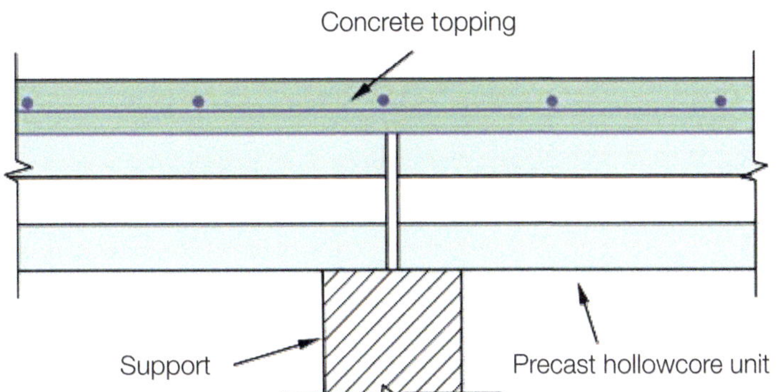

Not suitable for Consequence Class 2b and Consequence Class 3 buildings in accordance with Approved Document A[25]

This detail is acceptable for effective anchorage, but not to act as a horizontal tie unless mechanically connected to the planks.

Figure 10.4: Internal floor ties within hollowcore units

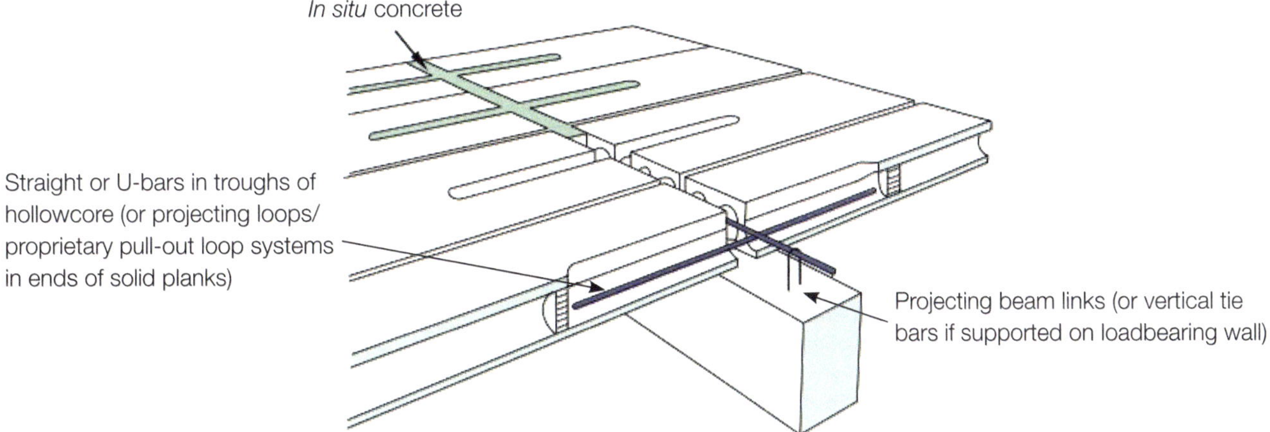

Figure 10.5: Position of floor ties for hollowcore units

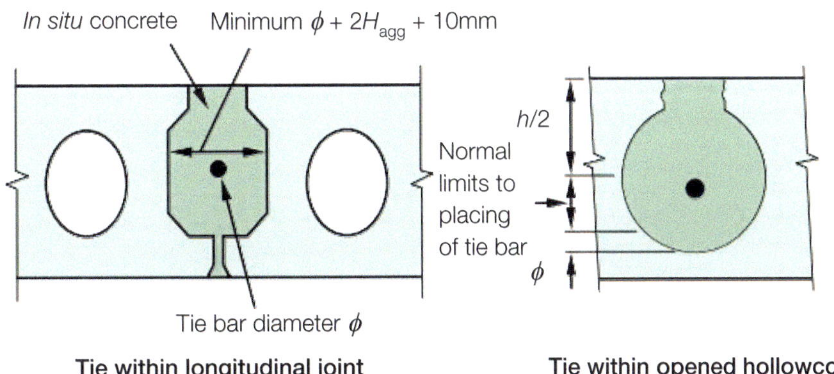

Tie within longitudinal joint

Tie within opened hollowcore

Figure 10.6: Interconnection between precast units onto an internal support beam

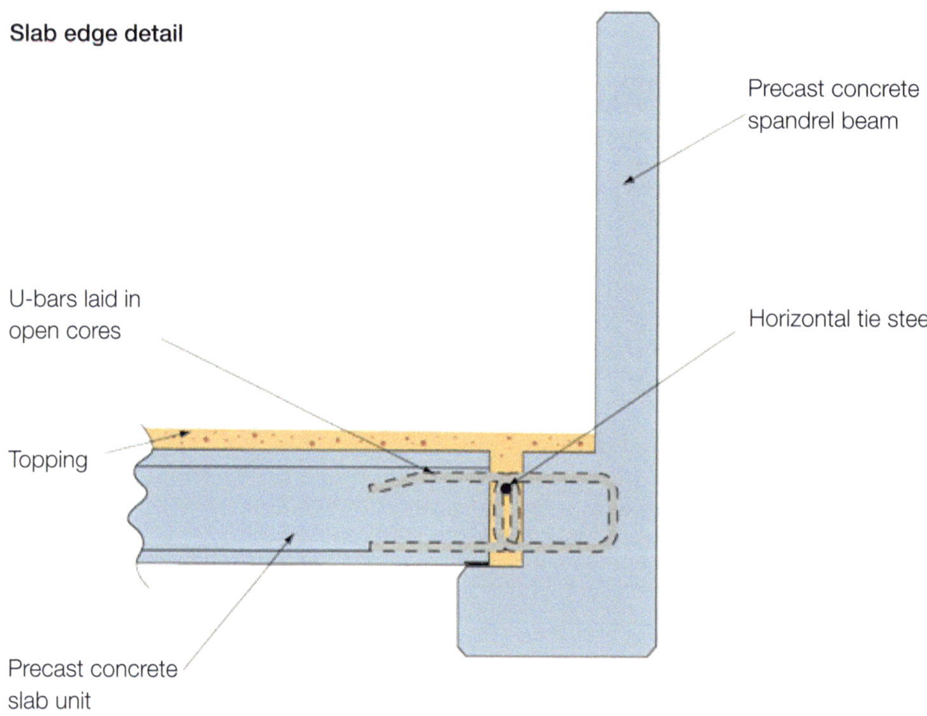

Figure 10.7: Interconnection between precast units onto an edge beam

Figure 10.8: Perimeter floor ties within hollowcore units

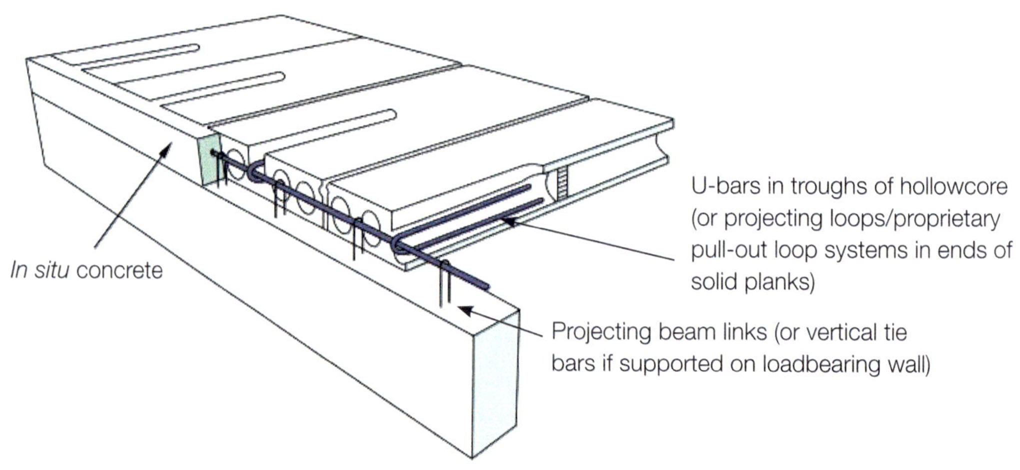

Figure 10.9: Perimeter ties where hollowcore units span parallel to edge beams

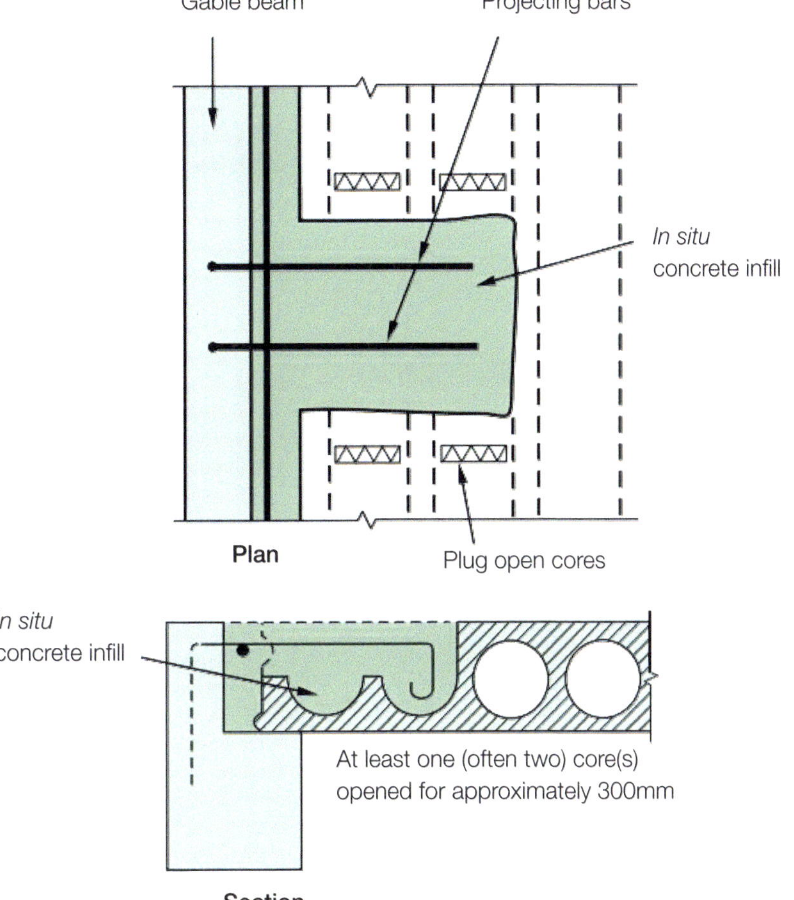

Figure 10.10: Interconnection of flooring units to non-loadbearing walls — hollowcore units span parallel to walls

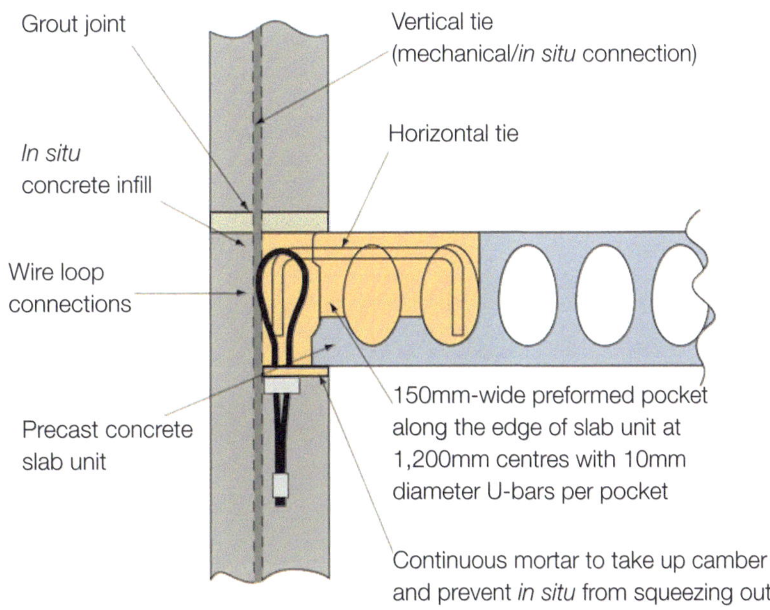

Figure 10.11: Detailing in corner columns

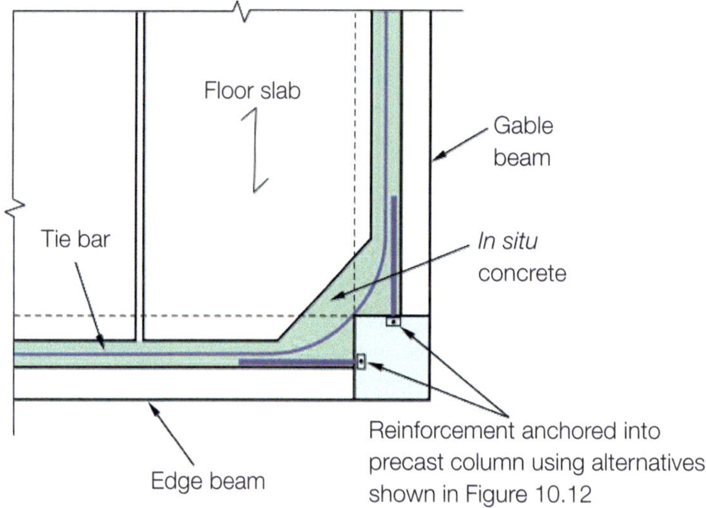

Figure 10.12: Alternatives for ties to a precast column

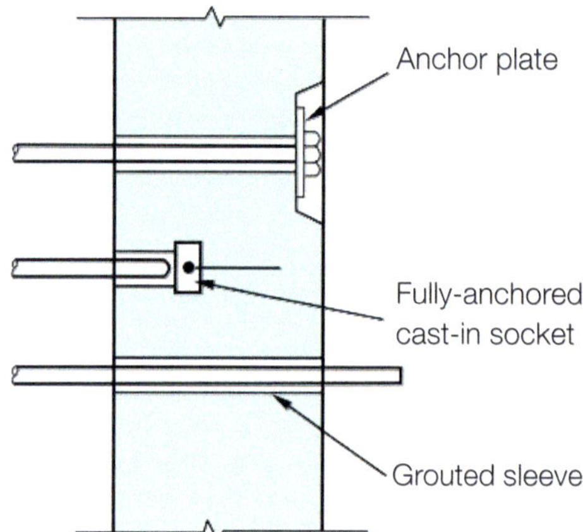

Figure 10.13: Internal ties taken through precast column

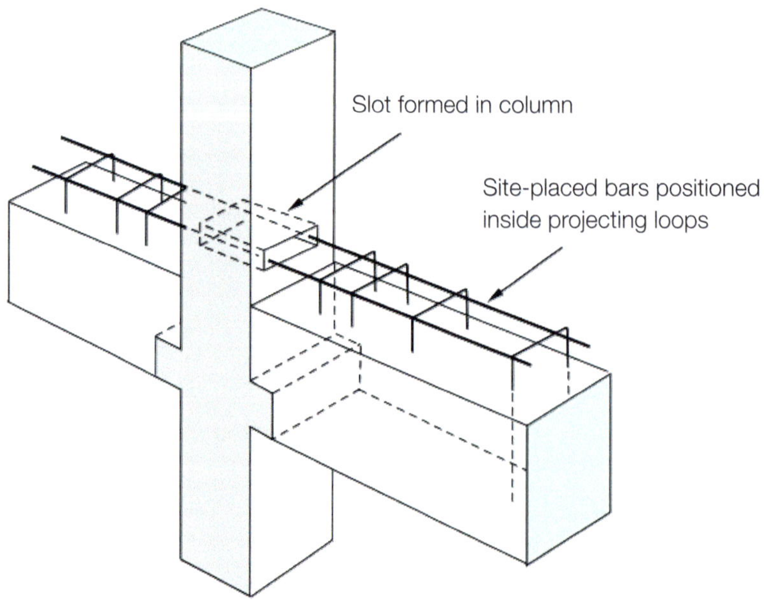

Figure 10.14: Ties for crosswall construction

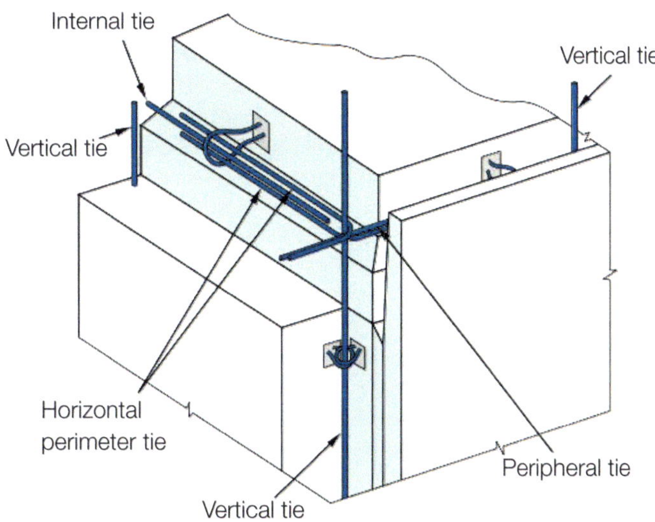

Case study 9

Precast concrete stair core — check of multiple approaches for robustness

Consider a 6m x 3m stair core in the ten-storey office building from Case study 7 (Figures 9.7 and 10.15).

Figure 10.15: Precast stair core

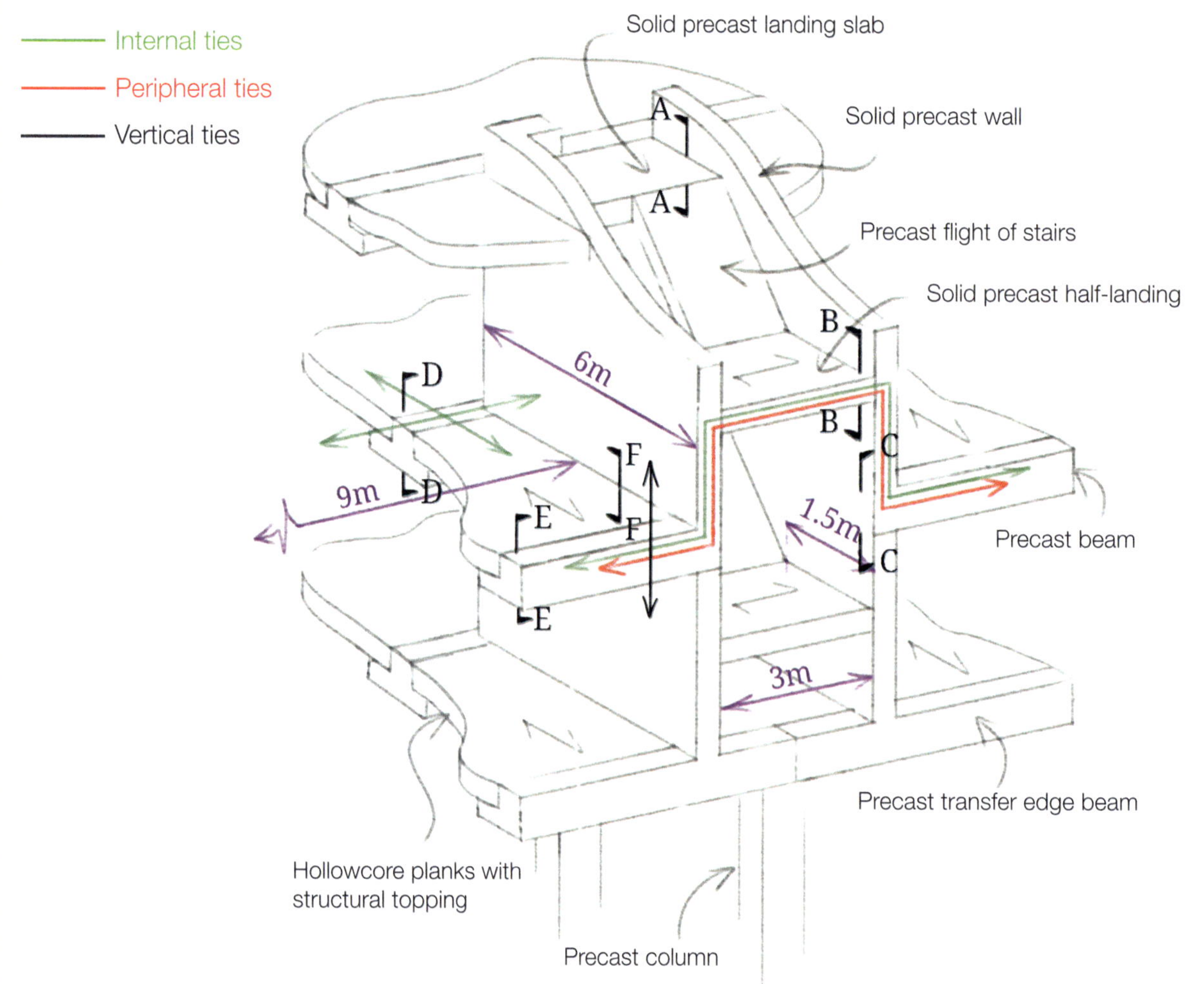

If the structure was converted to precast concrete with hollowcore planks spanning onto precast boot beams, supported on solid precast walls of the three-sided stair core as shown in Fig. 10.15, the structure would remain in Consequence Class 2b, and assuming, for convenience, the self-weight of the planks plus topping remains the same as the slab, the calculated tie forces would still apply.

Assuming that the core is supported on precast transfer beams at first-floor level, the design route is highlighted in Figure 10.16.

Note that this does not represent a particularly robust solution, but is chosen to demonstrate the different design approaches possible. The designer of such a building should advocate for the continuity of the vertical structure all the way to ground floor.

Figure 10.16: Decision routes followed in case study

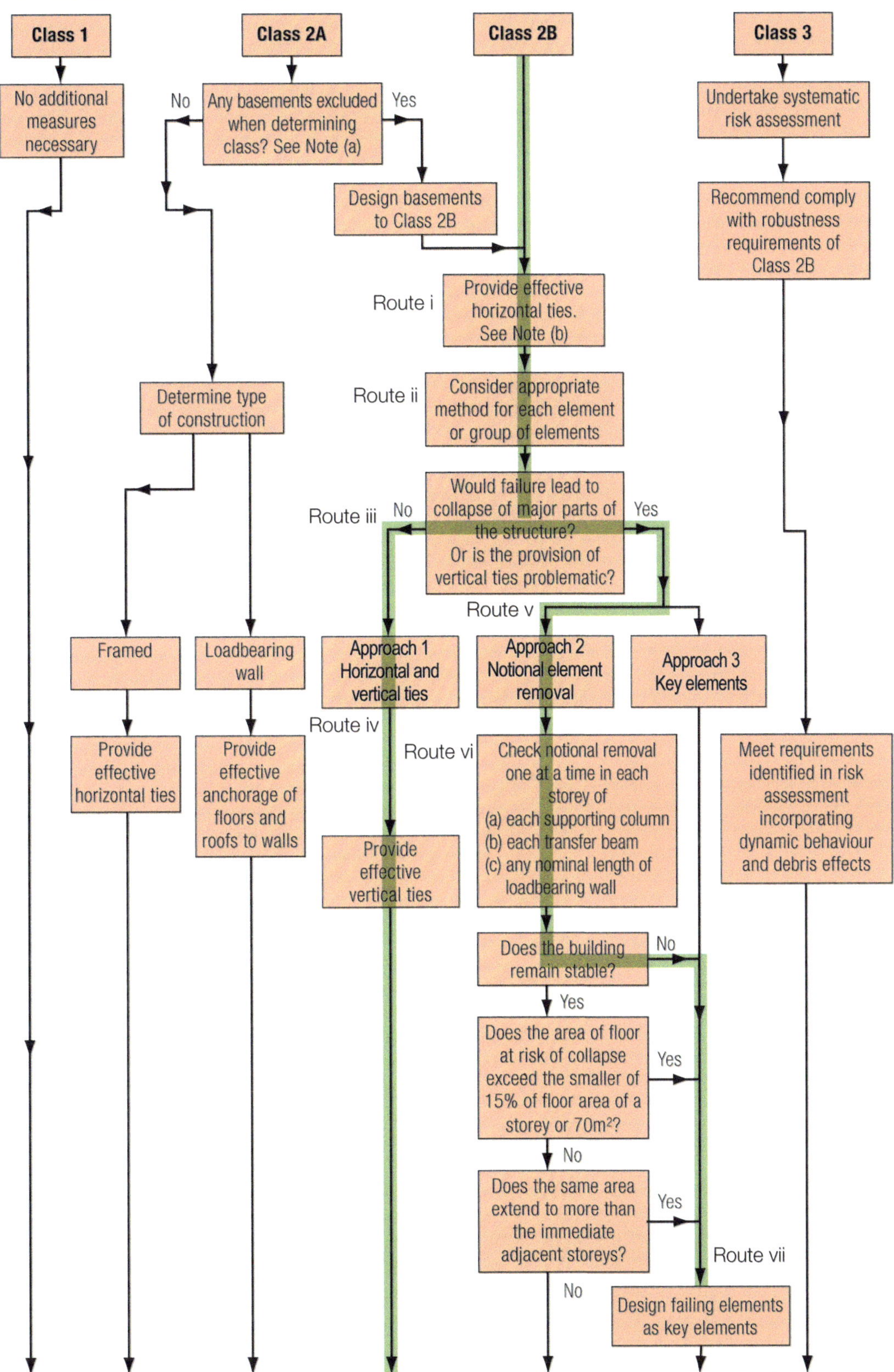

Calculation of tie capacities

Tie capacity $= A_s \times 500\text{N/mm}^2/1.0$

$1 \times$ B12 = 56.6kN

$1 \times$ B16 = 100.5kN

$1 \times$ B32 = 402.2kN

Firstly ensure the anchorage of any elements not acting as a tie is sufficient and anchored back to the part containing the ties.

Anchorages

Stairs to landing
Each end connection must be anchored to the landing with a force equal to dead weight, e.g., 30kN.

Design fixing for shear, or projecting bars into topping.

2 x B12 projecting bars provide tie force = 56.6kN x 2 = 113.2kN   OK

Landing to walls
Each end connection must be anchored with a force equal to dead weight of landing plus stairs, e.g., 20kN + 30kN/2 = 35kN (half stair flight to each end of landing).

Design projecting bars (or couplers) into slab topping (Figure 10.17).

2 x B12 provide tie force = 56.6kN x 2 = 113.2kN   OK

Figure 10.17: Section A–A — connection of landing to walls

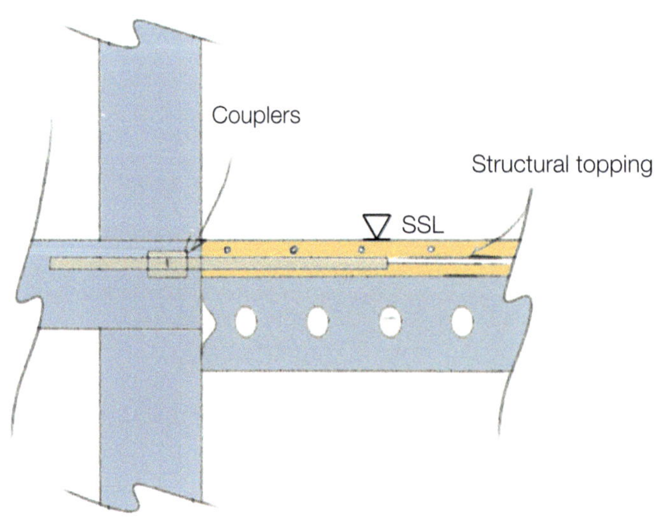

Checking Route i in Figure 10.16

Provide effective horizontal ties.

Half-landing to wall
The precast half-landing slab needs to act as a peripheral tie = 60kN (from Case study 7).

However, as ties are not possible through most of the stairwell due to the stairs, the half-landing will also need to resist the internal tie force of half a bay, concentrated to the edge of the building which would usually govern. This gives the opportunity for catenary action to occur.

Internal tie force = 189kN/m (from Case study 7) x 6m/2 = 567kN

A standard bearing angle with welded dowels would be unlikely to have sufficient capacity, therefore a structural topping should be provided, with pull-out bars from the precast wall (Figure 10.18).

B16s @ 200 centres provide tie capacity of 100.5kN x 1.5m/0.2m = 754kN   Therefore, OK

Figure 10.18: Section B–B — tying of half-landing to walls

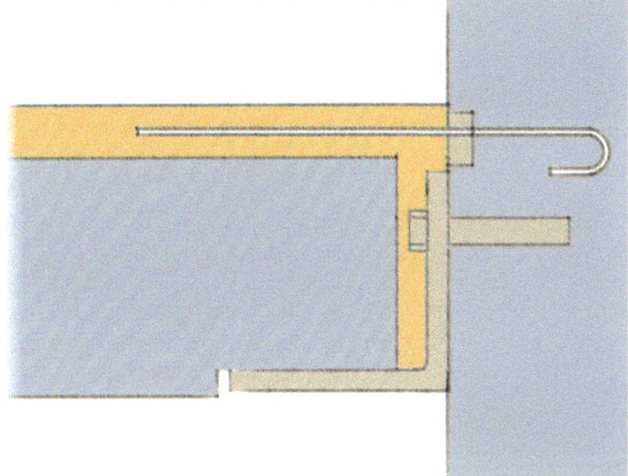

The most robust solution (where having walls on all four sides of the stair core is not possible) would be to continue a perimeter beam at main slab level to provide aligned continuity of tie forces. The least robust approach would be to treat the stairwell as a full void (cut-out in the floorplate) in which case the half-landing would only need to be anchored for its dead load (with the peripheral ties following the three sides of the core) (Figure 10.19). Using the half-landing as a perimeter tie is an intermediate solution between these two approaches.

Figure 10.19: Alternatives to using the half-landing as a perimeter tie

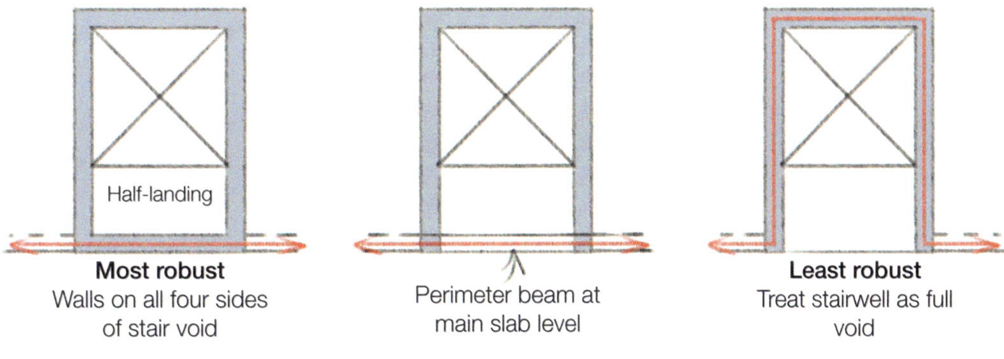

Edge beam to wall (Figure 10.20)

As given for the half-landing to wall example, internal tie force = 567kN

Shear capacity of B32 = 400kN x 0.6 = 240kN

Capacity of two dowel bars anchored into upper and lower wall = 4 x 240kN = 960kN   OK

Figure 10.20: Section C–C — connection of edge beam to wall

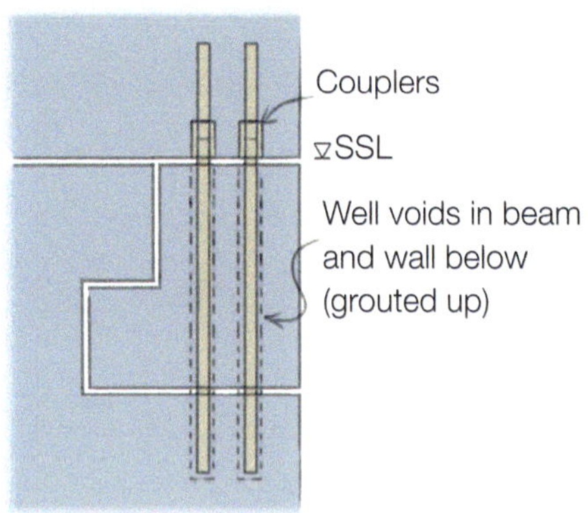

Plank to plank (Figure 10.21)

Internal tie force in the direction of the plank span (6m) = 126kN/m (from Case study 7).

Ties placed in open ends.

Use B12s to ensure full non-contact lap with strands in a max. 600mm-long open core.

Each U-bar has capacity 56.6 x 2 = 113.2kN.

Therefore, 2 x B12 U-bars per plank are needed (achieves 187kN/m).

Figure 10.21: Section D–D — tying the planks together

Central beam detail

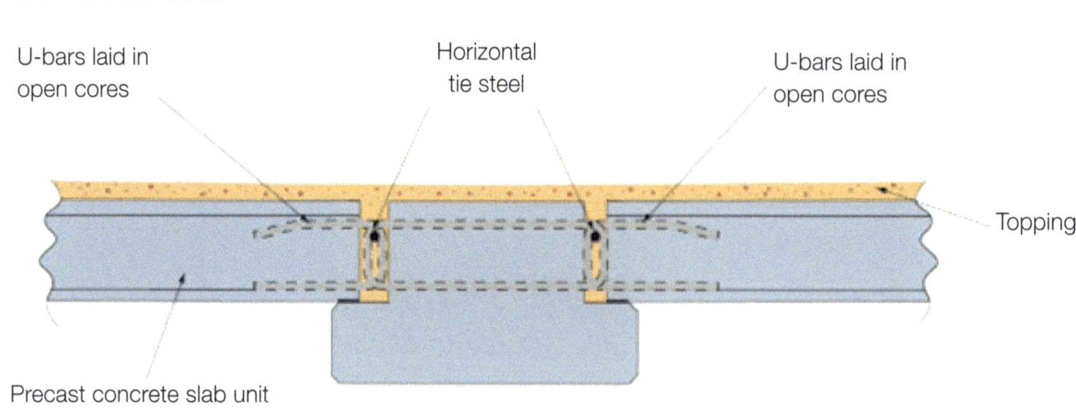

Plank to edge beam (Figure 10.22)
As given for the plank to plank example, internal tie force = 126kN/m, therefore 2 x B12 U-bars per plank are required.

Figure 10.22: Section E–E — tying plank to edge beam

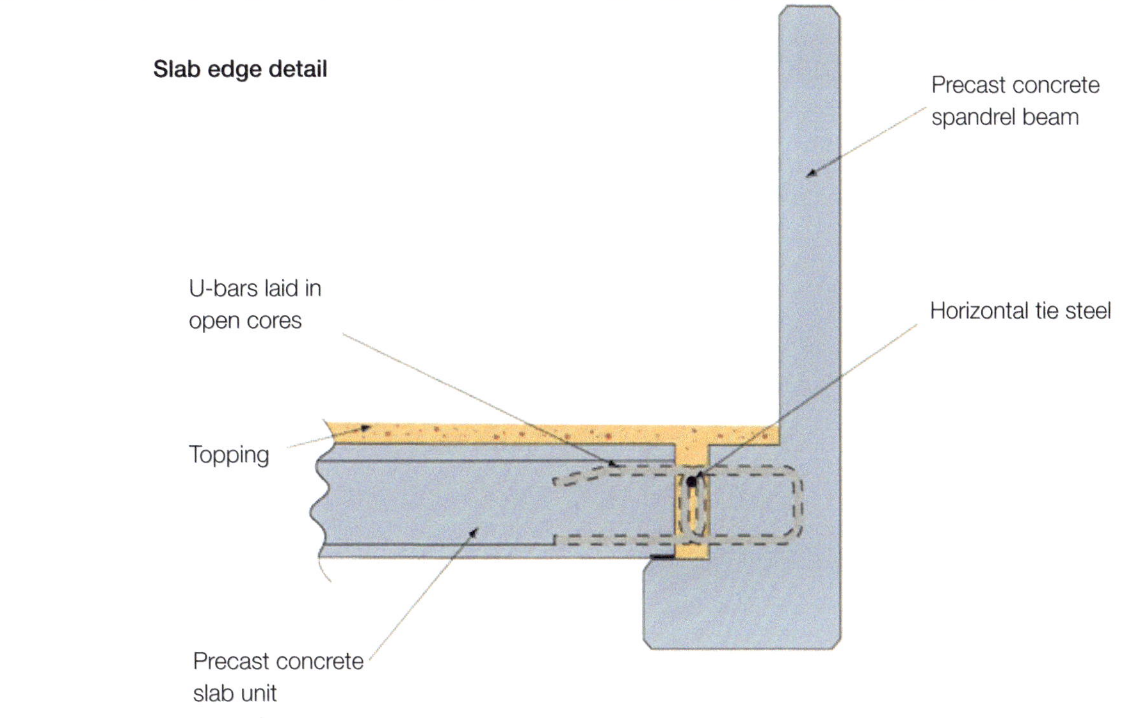

Internal beam to wall
Internal tie force in the direction of the beam span (9m): 189kN/m (from Case study 7) x 6m = 1,134kN (concentrated to beam).

Provide 4 x B32 continuous past edge of core, either projecting bars (or couplers) in beam end or in topping (achieves 350kN x 4 = 1,400kN tie capacity).

There are no external columns or walls in this part of the structure, which would otherwise need tying back into the floor slab.

Checking Route ii in Figure 10.16

Consider appropriate approach for each element:

Flanking wall in aligned structure above first floor
Failure of a flanking wall would not lead to collapse of major parts of the structure, and the provision of vertical ties is not problematic, therefore effective vertical ties should be provided (Route iv) (Figure 10.23):

Accidental load combination = 10.63kN/m^2 (from Case study 7).

Conservatively assume the same loading within the stair core:

Vertical tie force = 10.63kN/m^2 x (9m/2 + 3m/2) = 64kN/m

B16 dowels @ 1m centres provides 87.4kN/m   OK

Figure 10.23: Section F–F — vertical ties in wall

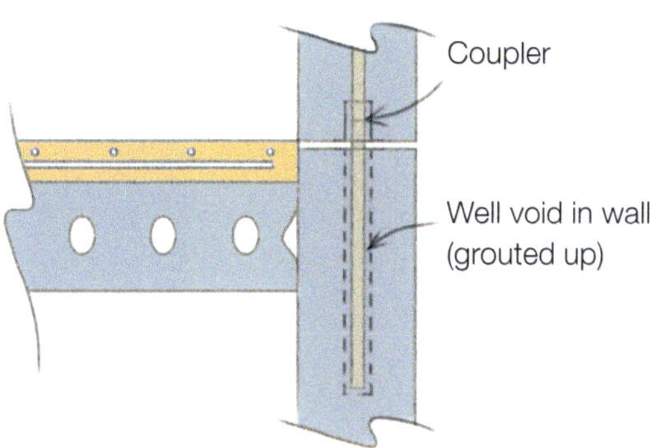

Removal of most vertical elements in vertically-aligned structures would not lead to collapse of major parts of the structure due to hanging and catenary action enabled through vertical and horizontal ties respectively. However, in the case of the flanking wall to the stair, the horizontal tie is not aligned (at the half-landing). Therefore, the flanking wall would need to be explicitly checked for: the out-of-plane bending of the wall spanning vertically between floors; the calculated tie force (in combination with the vertical load in the wall), and any second-order effects albeit with accidental load factors (Figure 10.24). If this is not feasible, the half-landing should be ignored in the robustness case and the walls treated as external walls (e.g., around a cut-out of the floorplate).

Figure 10.24: Before and after loss of a flanking wall leading to forces being induced in opposite flanking wall

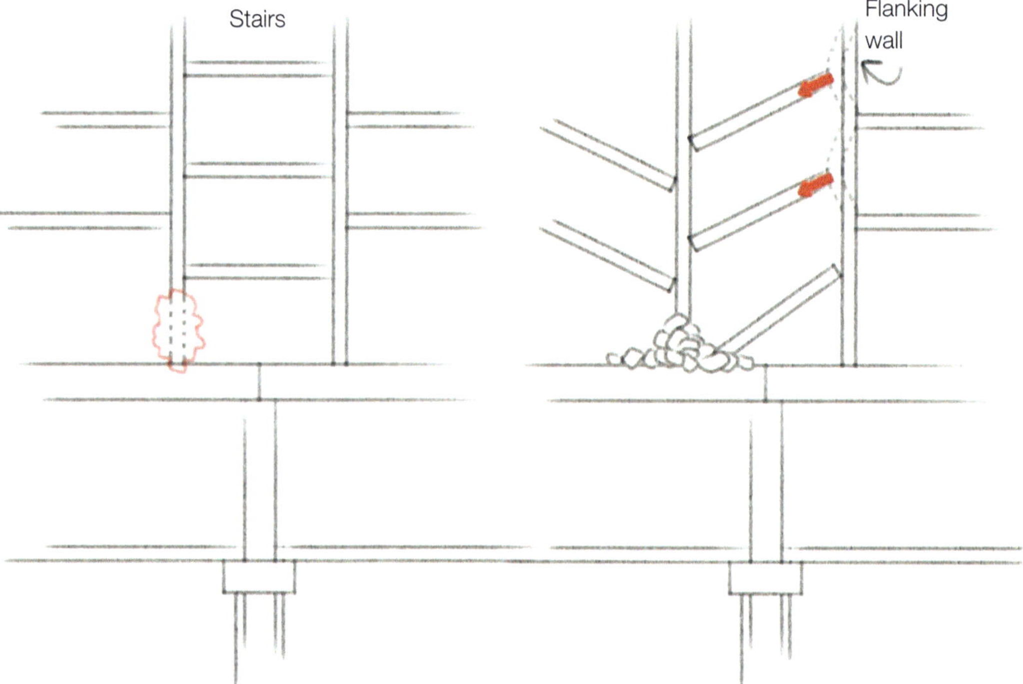

Checking other routes in Figure 10.16

Consider appropriate approach for each element:

Column supporting first-floor transfer structure
Route iii
Failure of this column would likely lead to collapse of major parts of the structure because the entire stair core would be unsupported, and catenary action unlikely to be possible.

Route v — Approach 2 — notional element removal:
The building does not remain stable if this element is removed (assuming the core is part of the stability system for the sake of this example, although in reality this would not be considered a robust design).

Route vii — Approach 3 — key element design:
Refer to Case study 8.

 *Design of precast concrete structures against accidental actions*[77]

 *Achieving robustness of precast concrete stairs using proprietary cast-in inserts*[78]

 *How to design concrete buildings to satisfy disproportionate collapse requirements*[79]

 *Robustness and disproportionate collapse*[80]

11 Design of masonry buildings

11.1 Introduction to masonry buildings

Masonry is the most common building material in the UK, with the majority of historic and modern homes being built in masonry. Prior to the development of steel and reinforced concrete, multi-storey buildings were also built in masonry and the UK's cities are full of medium-rise masonry buildings, built to celebrate and promote the companies that occupied them (Figure 11.1).

Figure 11.1: Industrial building built with brick external walls

A traditional masonry building, with thick, solid walls and a cellular layout is inherently robust (Figure 11.2).

Figure 11.2: Traditional masonry building

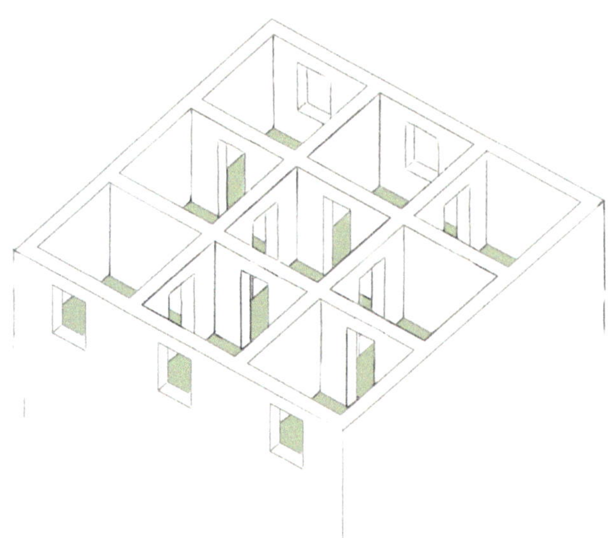

Masonry is the chosen material to create arched structures and, therefore, if a wall is lost in an accidental event, it can be expected that the remaining masonry will arch over the missing wall. Modern masonry buildings have larger openings, thinner and/or cavity walls, more penetrations and cavity trays and damp-proof courses throughout. They are very different from a traditional masonry building and may be less robust, and the arching properties of masonry may be compromised. However, research by CERAM (now Lucideon) to inform Approved Document A demonstrated that modern domestic-scale buildings are still very robust[81]. The trend for open-plan spaces and the move away from repetitive plan forms has caused engineers to avoid using masonry in high-rise construction.

11.2 Summary of requirements for each consequence class

Eurocode 6 contains no specific requirements for robustness, so it is necessary to refer to Annex A in BS EN 1991-1-7[4] and Table 19 of PD 6697[41] (Table 11.1 in this *Guidance*). The UK National Annex to Eurocode 6 references the BDA/APA/CBA Technical Note first published in 2005 and updated in 2013[82].

Table 11.1: Specific consequence class requirements for masonry

Consequence class		
1	**2a**	**2b**
Provided a building has been designed and constructed in accordance with the rules given in EN 1990 to EN 1999 for satisfying stability in normal use, no further specific consideration is necessary with regard to accidental actions from unidentified causes (A.4)	The provision of effective horizontal ties, or effective anchorage of suspended floors to walls, as defined in A.5.2 for buildings of loadbearing wall construction (A.4)	The provision of horizontal ties, as defined in A.5.2 for buildings of loadbearing wall construction together with vertical ties, as defined in A.6 in all supporting columns and walls should be provided **Alternatively** The building shall be checked to ensure that upon the notional removal of any nominal section of loadbearing wall, as defined in A.7, the building remains stable and any local damage does not exceed a certain limit Where the notional removal of such sections of walls should result in an extent of damage in excess of the agreed limit, such elements should be designed as key elements In the case of buildings of loadbearing wall construction, the notional removal of a section of wall, one at a time, is likely to be the most practical strategy to adopt
	Appropriate robustness should be provided by adopting a cellular form of construction designed to facilitate interaction of all components, including an appropriate means of anchoring the floor to the walls (A.5.2)	Continuous horizontal ties should be provided in the floors. These should be internal ties distributed throughout the floors in both orthogonal directions, and peripheral ties extending around the perimeter of the floor slabs within a 1.2m width of the slab (A.5.2) For horizontal ties to be successfully incorporated in a floor, a precast floor with screed or *in situ* concrete floor are the most practical ways

Note:
A.4–A.7 are all references to paragraphs in BS EN 1991-1-7

There is some confusion over whether horizontal ties need to be provided, if the approach taken is that of notional element removal. The answer to this is that they do (Table 5.1).

The word 'tie' is used a lot in building design, especially in masonry structures. The ties required and designed for robustness should not be confused with cavity wall ties, which connect two skins of masonry together at regular centres.

For CC2a buildings, the requirement is effective anchorage or horizontal ties.

Effective anchorage is indisputably less robust than effective horizontal ties. Effective anchorage is a bespoke solution for masonry (and timber). The authors of the Department for Communities and Local Government report[45] are very uncomfortable with effective anchorage being used instead of horizontal ties, but masonry buildings have been built and successfully inhabited for centuries. It is suggested that the engineer uses sound judgement in design; if a new CC2a masonry building has a traditional, cellular and regular layout (a robust layout), then effective anchorage may be sufficient. If the building has large open-plan spaces and irregular floor layouts, effective anchorage may not be appropriate.

11.3 Code requirements and other resources

11.3.1 Approved Document A
CC1 buildings include single-occupancy houses of four storeys or fewer, but Approved Document A[25] creates sub-divisions within this consequence class (Figure 11.3).

Figure 11.3: Spacing of horizontal restraint ties for Consequence Class 1 and 2a buildings

Construction	Spacing of horizontal restraint ties		
	Up to 3 storeys	4 storeys	5 storeys
Houses of Consequence Class 1 and 2a	2.0m†*	1.25m	1.25m
Other residential buildings of Consequence Class 2a	1.25m	1.25m	Not applicable

† When a timber joisted floor directly bears at least 90mm onto a supporting wall, no horizontal restraint ties are required from floor to wall (only applies to two-storey houses, designed to Approved Document A Section 2C, or to BS 8103-1, or to BS EN 1996).
* When a concrete floor directly bears at least 90mm onto a supporting wall, no horizontal restraint ties are required from floor to wall (applies to three-storey houses, designed to Approved Document A Section 2C, or to BS 8103-1, or to BS EN 1996).

It allows more leniency on providing horizontal restraint straps if a single-occupancy house is only three storeys with concrete floors, or two storeys with timber floors; it states that horizontal restraint straps can be omitted, subject to certain conditions.

This is consistent with the requirement for CC1 buildings stated in Table 11.1, where effective anchorage or horizontal ties are not mentioned.

In simple masonry structures, restraint straps are a very traditional part of cavity wall design, consisting of galvanised straps screwed to floors and wrapped over the inner leaf of a cavity wall. One of their key purposes is to provide restraint to masonry walls to stop them bowing out or failing under vertical or horizontal loads. They may help with robustness as they act as a form of horizontal tie. However, no calculation is required to determine the capacity of a restraint strap in a CC2a building or to check it against the horizontal tie formulae. They are also not continuous across the building.

11.3.2 BS EN 1991-1-7

Little more than what has been shown in Table 11.1 is included in this code for masonry buildings. It does provide the formulae for calculating horizontal and vertical tie forces for buildings of loadbearing wall construction, and in A.8 states the force for which a key element should be designed; that of $34kN/m^2$.

BS EN 1996-1-1 (Eurocode 6)[40], which is the design code for masonry, does not provide guidance on designing for robustness, and therefore the design engineer must refer to PD 6697[41].

11.3.3 PD 6697

PD 6697 is entitled '*Recommendations for design of masonry structures to BS EN 1996-1-1 and BS EN 1996-2*'.

It is a document published in the UK that contains complimentary, non-contradictory information to support Eurocode 6. It is similar to the earlier (now withdrawn) BS 5628-1[83]. The relevant section is 6.5. PD 6697 is essential if designing a masonry building, because it contains much more specific detail than BS EN 1991-1-7.

Within Section 6.5, Table 18 (adapted as Table 11.2 in the *Guidance*) defines loadbearing elements (either walls, columns, beams or slabs). This definition is required if it is decided to take the notional element removal approach, to check the building for disproportionate collapse. The table also introduces the concept of lateral support, which is further defined in Section 6.5.6.

There is a tendency to consider loadbearing elements as only walls or columns but — as is the case for all materials — beams can also be significant loadbearing elements if they support columns or walls.

Table 11.2: Definitions of loadbearing elements

Type of loadbearing element	Extent
Beam or slab supporting one or more columns or a loadbearing wall	Clear span between supports or between a support and the extremity of a member
Column	Clear height between horizontal lateral supports
Wall incorporating one or more lateral supports	Length between vertical lateral supports or length between a vertical lateral support and the end of the wall
Wall without lateral supports	Length not exceeding $2.25h$ anywhere along the wall (for internal walls) Full length (for external walls)

Notes:
Lateral supports to walls are provided by intersecting or return walls, piers, stiffened sections of wall or substantial loadbearing partitions or purpose-designed structural elements.
Under accidental loading conditions, temporary supports to slabs can be provided by substantial or other adequate partitions capable of carrying the required load.

In summary, when an engineer is taking the notional element removal approach, the notional length of wall removed is between lateral supports, or $2.25h$ if there are no lateral supports to the wall, unless it is an external wall. If the wall is an external wall, the length to be removed is the full length of the wall.

This onerous requirement for an external wall emphasises the importance of having a cellular layout; it is unlikely much of a building would survive the removal of an entire external wall.

Lateral supports are described in Section 6.5.6 of PD 6697. It is important to consult this section before finalising a building plan, because lateral supports need specific capacities of connection to the loadbearing wall, and also have prescribed lengths. For example, if it is a return wall providing lateral support, it needs to be at least $h/2$ in length.

The rest of Section 6.5.6 includes the formulae for designing horizontal and vertical ties and illustrates how horizontal ties should be placed. There is also the implication that CC2b buildings should have precast or *in situ* concrete floors. The section also includes the calculation for the capacity of masonry to resist the horizontal pressure force set out in Section 6.5.2.

11.3.4 Effective anchorage

The term 'effective anchorage' is generic and could mean any of the following:

- Straps or restraint-type joist hangers with timber joists
- A precast concrete plank floor, precast concrete beam and block floor or *in situ* concrete floor spanning onto a wall, or where the precast floors run parallel to the wall, built into the wall
- A precast concrete floor running parallel to the wall, provided with restraint straps

Effective anchorage is an integral part of the design of most simple masonry buildings.

If an older building is undergoing structural alterations, it may be discovered that the connection of floors to walls are not sufficient to be described as effective anchorage, so engineers frequently create pockets in the masonry and cast restraint straps into these pockets, and fix the straps to the floors.

Although most effective anchorage details do act as ties to some extent, they are not subject to the more onerous calculation of a true horizontal tie and will not have the same capacity.

Precast flooring units usually span in one direction and shed no load onto walls parallel to their span. It is therefore unsafe to assume that any such walls below will be restrained by friction against out-of-plane movement unless bedding mortar/positive anchorage is provided.

11.3.5 Horizontal ties

Where horizontal ties are required, calculation is necessary and the horizontal tie force requirements are set out in A.5.2 of BS EN 1991-1-7, Expressions A.3 and A.4 for internal ties and peripheral ties respectively, and Table 20 of PD 6697 where greater detail is given.

For internal ties T_i = the greater of F_t kN/m or $\dfrac{F_t(g_k + \Psi q_k)}{7.5} \dfrac{z}{5}$ kN/m   (Expression A.3)

For peripheral ties $T_p = F_t$ (Expression A.4)

Where:
F_t =   60kN/m or 20 + 4n_o kN/m, whichever is less
n_o =   total number of storeys in a structure
z =   the lesser of:
- five times the clear storey height H, or
- the greatest distance in metres in the direction of the tie, between the centres of the columns or other vertical loadbearing members whether this distance is spanned by:
  - a single slab, or
  - a system of beams and slabs

The capacity of the tie force for the tie that connects floor to wall is not defined in BS EN 1991-1-7, so the design engineer must consult PD 6697 (Table 20 in particular). This tie force is calculated by the expression:

$2F_t$ or $(h/2.5)F_t$, whichever is the lesser; where h is measured in metres and is the clear storey height

PD 6697 uses L_a instead of z, but L_a is defined in exactly the same way.

The *Structural masonry designers' manual*[84] demonstrates in an example calculation, that the tie connection of the slab to the wall may be achievable with only the shear capacity of the mortar joint. However, even if a tie can be provided, its natural tendency will be to 'rip away' the masonry and, therefore, a mortar joint always needs to be checked.

There has been some suggestion that horizontal ties in the form of a rod across the whole building will enable a floor slab to span in catenary action across a missing wall. Tie rods are frequently seen in historic buildings, but the artisans who built them would probably not have been familiar with the term 'disproportionate collapse'.

11.3.6 Vertical ties

The calculation of vertical tie force is set out in Section 6.5.5 and Table 21 of PD 6697 and Section A.6., Expression A.5 in BS EN 1991-1-7.

$$T = \frac{34A}{8000}\left[\frac{H}{t}\right]^2 \text{ N, or 100kN/m of wall, whichever is greater} \qquad \text{(Expression A.5)}$$

Where:

A = the cross-sectional area in mm^2 of the wall measured on plan, excluding the non-loadbearing leaf of a cavity wall.

Within Section 6.5.5 it is implied that timber floors are not acceptable, if taking the approach of using vertical ties.

11.3.7 Designing structure as a key element

If an element needs to be designed as a key element because its removal would result in large-scale building collapse, it should be designed for a load of $34kN/m^2$. BS EN 1991-1-7 gives no guidance on how to check a masonry wall for this load, and again a design engineer needs to refer to PD 6697, Section 6.5.2.

The lateral strength of the masonry is calculated as:

$$Q_{lat,d} = \frac{7.6tN_{Ed}}{h^2}$$

Where:

N_{Ed} = the axial load on the wall.

To understand the lateral strength of a masonry wall, consider walls of various thicknesses (Table 11.3). The first wall could be an external wall of a Victorian property and the second wall possibly that of a church.

Table 11.3: Lateral strength of masonry walls of 200mm and 400mm thickness

Thickness of wall, t	200mm	400mm
Axial load on wall	50kN/m	50kN/m
Height of wall, h	3.4m	3m
$Q_{lat,d}$	$6.6kN/m^2$	$17kN/m^2$

It can be seen in these calculations that even a relatively thick, 400mm, wall cannot resist a load of $34kN/m^2$. Therefore, designing masonry elements as key elements may not be an appropriate approach, unless designing a building with very thick walls.

11.4 Masonry cladding

Today, apart from use in low-rise housing, the most common use of masonry is in cladding for buildings constructed from other materials such as lightweight steel frames, timber frame or concrete and steel frame (Figure 11.4).

Historically, masonry used as cladding was provided with horizontal movement joints every few storeys, but there are economic reasons for building the entire facade with no interruption.

Figure 11.4: Masonry cladding on framed building

SCI Technical Information Sheet P426[85] discusses various issues to be considered when taking this approach, and gives advice on the disproportionate collapse of the cladding. After all, even though the masonry is simply cladding, large-scale collapse could still be very dangerous (Figure 11.5).

Figure 11.5: Oxgangs Primary School facade collapse

After the Oxgangs Primary School, Edinburgh facade collapse the IStructE produced *An overview of the specifying and detailing of masonry construction*[86] which concluded that properly-designed facades are not at risk of collapse.

11.4.1 Masonry on support angles

In multi-storey steel- or concrete-framed buildings, masonry is often used as cladding and will be supported on brackets fixed to the principal structure. This is generally designed by the bracket manufacturer, and if correctly installed poses no particular issues. Co-ordination between engineer, architect and manufacturer is critical to ensure correct installation. Brackets are typically provided at every other floor, but if there are many openings, the weight of the masonry will be concentrated down facade columns, and it is becoming increasingly popular to provide brackets at every floor level.

 *Structural masonry designers' manual*[84]

11.5 Masonry building options

Although it has been stated that design engineers are avoiding using masonry to design multi-storey buildings, there are some who are finding ways to satisfy the robustness requirements of a CC2b building with a structure of masonry construction.

Case studies 10–12 look at three different approaches. Case study 10 covers what was formerly the most common approach, that of constructing a masonry building over a reinforced concrete podium. This was done frequently on large developments but only on buildings of five storeys, and was supported by a guidance document produced by NHBC[52], although it is an approach criticised in the Department for Communities and Local Government report[45]. It is recommended that this approach is no longer followed (Section 6.2.2).

The second approach is to incorporate horizontal ties but, rather than provide vertical ties, look at removing wall elements (Case study 11).

The final approach is to provide the masonry building with a high number of lateral supports (Case study 12). Table 18 in PD 6697 contains a note stating that a lateral support can be provided by purpose-designed structural elements.

Case study 10

Masonry building over reinforced concrete podium

Important note: This example is of a form of design that is no longer considered acceptable. It is included to assist design engineers working on alterations to buildings that may have been designed in this way.

Large new developments were often a mix of low-rise houses and apartment blocks, with most apartment blocks being under five storeys and, therefore, being classed as CC2a buildings. However, it was common for the developer to want to provide the occasional five-storey building, and to construct it in a similar way to the other buildings on the site. This created a problem in that a five-storey residential building is classed as CC2b.

Historically it had become common practice on large developments to design five-storey buildings as four-storey CC2a buildings, sitting on a CC2b reinforced concrete structure (Figure 11.6). This was accepted by warranty providers such as NHBC and Premier Guarantee.

Figure 11.6: Historic approach to robustness of a five-storey building

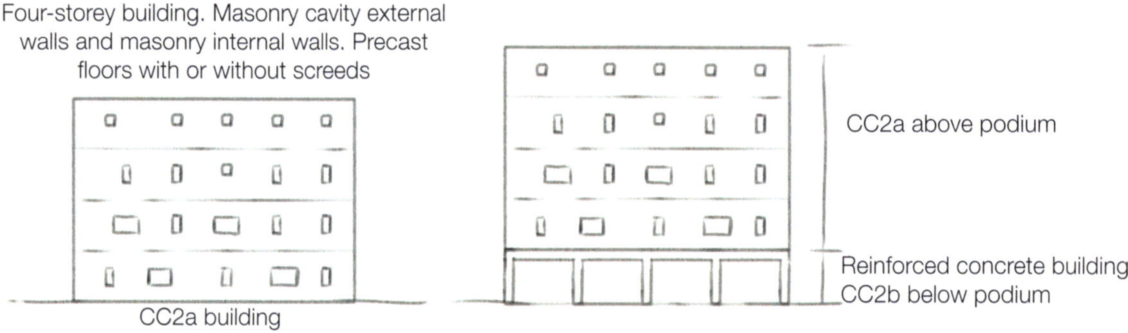

There was confusion over what forces the podium deck and ground-floor columns needed to be designed for, and engineers took a mix and match approach of:

* Designing for debris loading on the podium deck and designing the ground-floor columns as key elements
* Simplifying this approach by not carrying out a check on the debris weight, on the assumption that the permanent structure weighed the same as the debris, and the design for permanent loading (with factored loading) was the same or more onerous than designing for the debris, although applied as line loads and not a distributed load
* Designing the podium and ground-floor columns as key elements for a load of 34kN/m^2
* Following any one or none of these approaches

The most common approach seemed to be the second one, probably because it did not require further calculation. However, this approach was inappropriate because it failed to consider designing the podium and ground-floor columns (which are obviously very important) as key elements.

The correct approach would have been to design the podium deck and ground-floor columns as key elements, to sustain a load of 34kN/m^2 and to check the podium floor for debris loading.

Structural engineers working on alterations to buildings designed in this era will need to check what approach the original engineer took.

Case study 11

Loadbearing masonry building with reinforced concrete elements

The example described here is a simplified version of a building that the design engineer had not realised was a Consequence Class 2b building until design and procurement was complete. The design engineer was responsible for the structural engineering of four-storey apartment blocks across the site. These were all constructed in loadbearing masonry with precast concrete floors.

Only one of the buildings on the development was five storeys, and it had been decided that this would be built in the same way.

However, the elevation of building class from CC2a (the four-storey buildings) to CC2b (the five-storey building) meant that the robustness strategy needed to be reviewed and, after this was pointed out by the checking engineer, the design engineer revisited the design scheme.

The design engineer decided to take the second approach; that of removing wall elements between lateral supports to assess what would happen (Figure 11.7).

Figure 11.7: Plan of loadbearing masonry building with reinforced concrete elements

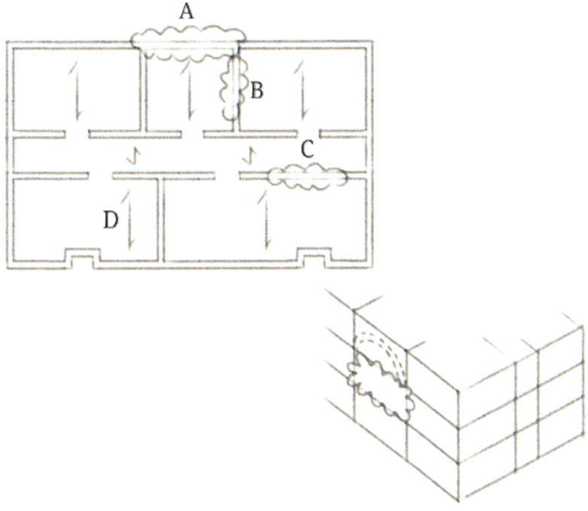

This example highlights four different locations of wall:

- The wall at A was supporting a floor and the inner skin of the masonry (Figure 11.8)
- The wall at B did not support a floor, only the internal walls above
- At C, the wall supported both a floor and an internal wall
- D was included to look at the performance of the floor if a supporting wall was removed

Figure 11.8: Detail at A

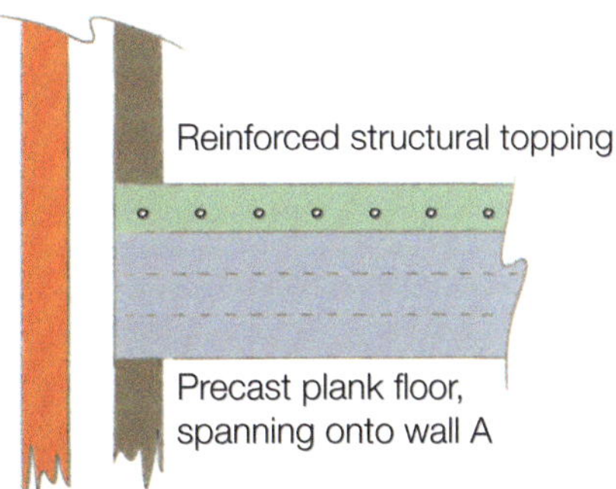

The floor was of precast planks with a reinforced structural topping. The reinforcement may enable a floor to span in the opposite direction to that of the planks.

At A, if the wall below were removed, the structural topping would need to be able to support not only itself but the inner skin of masonry above.

It was concluded that this was not possible, so an edge beam was inserted into the floor and topping (Figure 11.9). The engineer decided it was not necessary to check whether the masonry could arch.

Figure 11.9: Detail at A with reinforced concrete edge beam

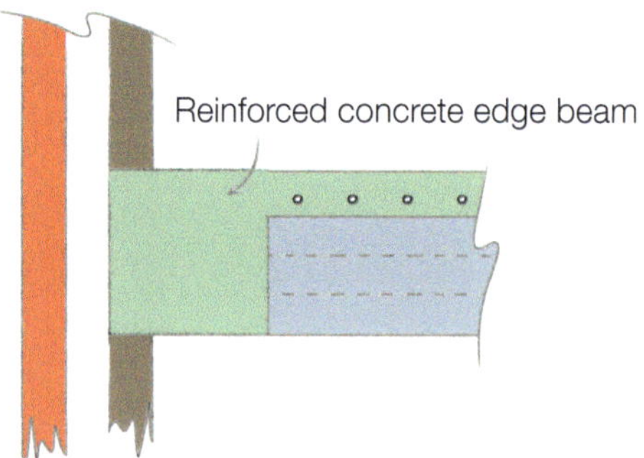

The wall at B was not supporting a floor. It was only supporting the wall above which, in itself, was not loadbearing. Therefore, it was decided that the removal of wall B would not be a problem; the wall above would be supported by the precast plank. If this was not successful, the loss of the wall above B would not create significant problems further up the structure.

At C, the problem was very similar to that of wall A, and therefore it was decided to put in a reinforced concrete beam integral with the structural topping.

Finally, it was checked that the structural topping could act as a floor, spanning in the opposite direction to the precast plank (at D) and, due to reductions in imposed load, it was concluded that it could.

The final scheme looked similar to Figure 11.10.

Figure 11.10: Reinforced concrete beam added to loadbearing masonry building

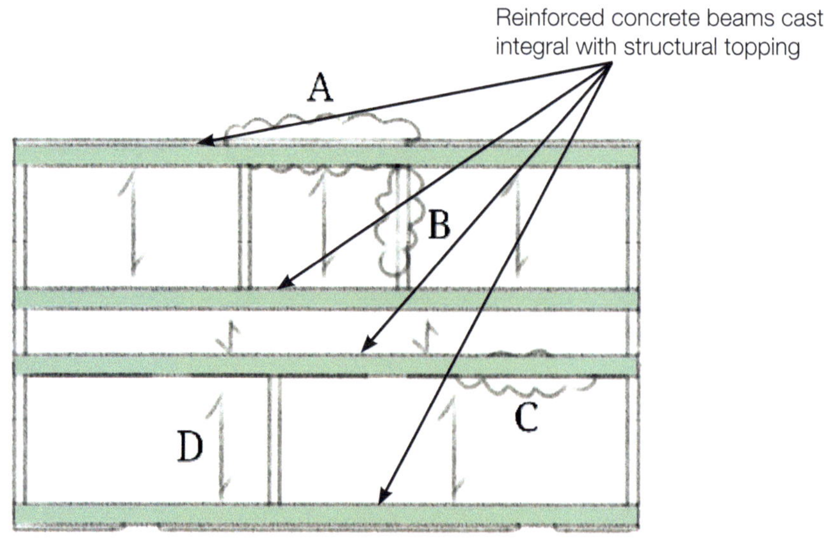

Case study 12

Loadbearing masonry building with reinforced concrete lateral supports

This masonry building example details the approach of providing lateral supports, and checking that the masonry walls between lateral supports can be removed one at a time.

The floors were of *in situ* concrete and therefore could easily span in the opposite direction if their support wall was removed.

However, the design engineer took the approach of using reinforced columns within the masonry as lateral supports, as opposed to using perpendicular walls (Figure 11.11). This is acceptable as Note 2 in Table 18 of PD 6697 states:

"Lateral supports can be provided by purpose-designed structural elements."

Figure 11.11: Lateral supports provided by purpose-designed structural elements

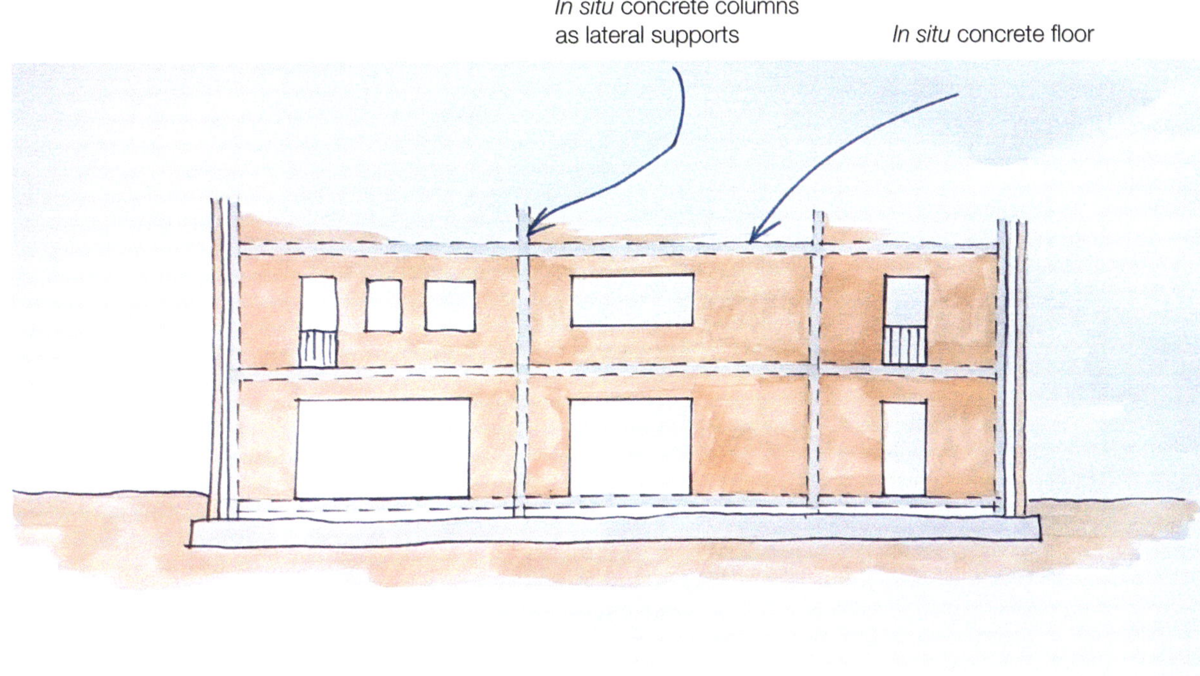

12 Design of timber buildings

12.1 Introduction to timber design

Timber is used for construction of buildings of all types. It is commonly used for new low-rise houses, and also for multi-storey buildings. Timber is used in masonry buildings within floor and roof structures. As a lightweight material it is a popular choice for constructing upper storeys of buildings over ground floors built with reinforced concrete or steel.

Traditionally, timber has been used in its natural form, sliced from trees into studs, beams and planks. Today it is frequently used in an engineered form, either in solid beams (glulam, laminated veneer lumber (LVL), etc.) or I joists, with an LVL or steel web.

In recent years, cross-laminated timber (CLT) has been used for solid walls and floors, either on its own or in conjunction with a steel frame. Finally, the stalwart of quick house construction on-site is the trussed rafter for roof construction.

Like other materials, a building's structural form will significantly affect its robustness. Cellular forms of construction such as those often seen in platform timber-frame buildings[87] with many loadbearing walls, provide a high level of robustness and resistance to accidental damage, because loss of any one wall will generally not lead to the collapse of a large section of the structure.

Essentially the same principles apply as for any other material. The first of these is that the inherent structural form should lend itself to assuring robustness, the second is that buildings should be designed for derived horizontal loads, and thirdly, all components need to be properly interconnected.

This chapter follows the requirements of BS EN 1995-1-1[42] and BS EN 1991-1-7[4] and the supplementary National Annexes[88,33]. There are no specific requirements for robustness in Eurocode 5 and, therefore, the engineer should refer to PD 6693-1[43]. When designing multi-storey timber-frame buildings it is recommended that reference is made to the paper on 'Verification of the robustness of a six-storey timber frame building' in *The Structural Engineer*[89].

Finally, the TF2000 tests by the British Research Establishment in Cardington[90] informed much of the subsequent guidance on timber building design, and these are also referred to in this chapter.

12.2 Summary of requirements for each consequence class

The methods for achieving robustness for CC2a will differ depending on the form of timber structure. The methods can be broadly described as either:

- Provision of horizontal ties, or
- Provision of effective anchorage of floors to walls for loadbearing wall forms of construction

There are three methods of achieving the requirements of Consequence Class 2b:

- Provision of both horizontal and vertical ties
- Notional removal of elements
- Provision of structurally adequate key elements

Again, the methods for achieving robustness for CC2b will differ depending on the form of timber structure. For example, mass timber buildings are not typically provided with discrete rim beams at floor perimeters, and therefore require a different approach to platform-frame buildings, comprising joisted floors with rim beams.

The solution adopted may be a combination of these approaches for different areas of the building, depending on the structural form. A description of the most popular structural forms of timber construction is provided in Section 12.3 and the most appropriate way to provide structural robustness of each form is described in Section 12.9.

Detailing of timber buildings requires attention to robustness at all stages of construction and at all material interfaces, since the way many timber structures are constructed relies on considerable interaction between primary and secondary members and mechanical fixings. Indeed, if the notional element removal or key element approaches are taken, it is expected that effective anchorage is also provided, and that there is adequate interconnection between vertical and horizontal structural elements.

12.3 Forms of timber construction

The most common structural forms available for timber buildings are:

- Trussed rafter roofs. In the UK, trussed rafters are commonly used within hybrid structures
- Platform timber-frame construction comprising loadbearing wall elements of timber studwork or structural insulated panels, constructed off the floor structure (or platform) below. Platform timber-frame construction is used for residential buildings up to seven storeys, but this may be limited by considerations of fire robustness (Section 12.10)
- Mass timber comprising CLT floor and wall panels, usually constructed as platform frame but may also be of balloon frame where floors are supported off the inside of the continuous wall panels
- Post and beam frames, typically with braced bays or discrete shear walls and using glulam beams for longer span structures, sometimes combined with CLT floors
- Portalised frames. Glulam portal frames have been used for large-span open-plan structures

There are specific considerations to achieve robustness for each of these structural forms.

12.4 Horizontal loads

12.4.1 Notional horizontal loads
Although not directly mentioned in BS EN 1995-1-1, the general Eurocode approach of accounting for column 'out of verticality' or notional inclination also results in the generation of horizontal loads. Wind loading is likely to be the source of dominant horizontal load. Nevertheless, each storey should also have sufficient strength and stiffness to resist a horizontal, long-term force. BRE Report BR454[90] recommends that 2.5% of the vertical load (defined as that percentage of the dead + imposed loads) should be applied as a horizontal load.

12.4.2 Diaphragm action for resistance to lateral loading
As part of the standard design process, suitable bracing in the form of either discrete bracing or braced wall diaphragms, should be provided in vertical structure. Horizontal diaphragms in the form of a structural subdeck to floor structures and roof structures, with directly-fixed ceilings, sarking or discrete bracing members should be provided in planes parallel to the direction of the lateral forces, acting on the whole structure.

For all timber structures it is necessary to check assumptions of diaphragm action to ensure the design is appropriate, and specific checking is required at each level to confirm that the transfer of horizontal forces is adequate. This should include checking of all horizontal-to-vertical diaphragm interfaces to verify that appropriate in-plane and lateral shear forces, arising from the design wind loading, can be transferred at each level.

BS EN 1995-1-1, Clause 9.2.4.3 and PD 6693-1, Clause 20 provide a simplified method for designing sheathed wall diaphragms in timber-frame buildings.

PD 6693-1, Annex E provides methods for ensuring adequate bracing to horizontal diaphragms comprising trussed rafters, and further guidance can be found in STA Advice Note 17[91] describing required connections between horizontal and vertical diaphragms at roof level.

Special care should be taken where wall panels are not braced by horizontal diaphragms at regular intervals (e.g., full height external panels in stair cores), in which case wind posts may be required.

Where the action of an efficient ceiling diaphragm cannot be guaranteed (e.g., where a decoupled resilient bar type ceiling is provided), Annex B of PD 6693-1 indicates that additional edge blockings should be provided at floor perimeters where joists run parallel to the wall, to ensure the lateral stability of the floor edge members (Figure 12.1).

Figure 12.1: Platform frame floor component interfaces

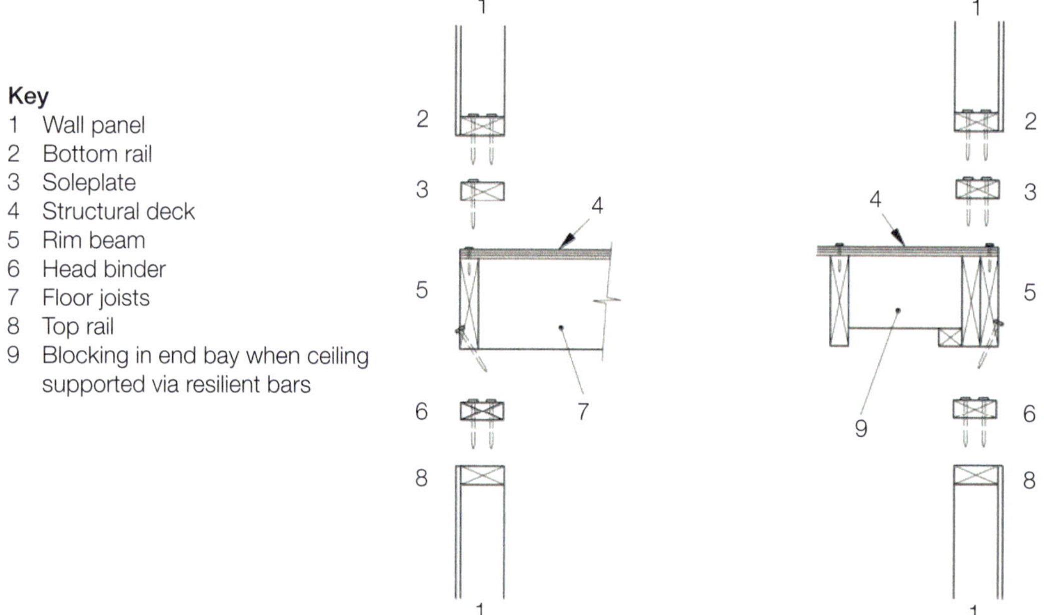

12.5 Effective anchorage in Consequence Class 2a buildings

For CC2a buildings, robustness will be achieved by providing effective horizontal ties, or effective anchorage of suspended floors to walls.

The tests on the TF2000 building at Cardington demonstrated that correctly-detailed timber-frame structures will possess a degree of horizontal tying by the fixing densities provided between elements for overall building stability, without the requirement for distinct horizontal ties.

The structural layout of platform timber-frame buildings is cellular in nature, and effective anchorage of each floor member to wall junction in accordance with Clause A.5.2 of BS EN 1991-1-7, Clause NA.2.43 of the UK National Annex and Annex B of PD 6693-1 is the preferred approach.

Effective anchorage of floors to walls will be achieved with a minimum density of nails in all horizontal interfaces, equivalent to 3.1mm diameter at a maximum spacing of 300mm centres as shown in Fig. 12.1. The density of fixings has been derived from test evidence backed by calculations undertaken by the Structural Timber Association (STA, formerly the UK Timber Frame Association (UKTFA)), and submitted to the UK code committee for approval and publication in PD 6693-1. The density of fixings gives an appropriate uniform distributed lateral tie force. Distinct horizontal and vertical ties are not required to be provided.

In all cases, the design process should involve checking the capacity of the component interfaces along the load path (e.g., panel rail to soleplate, soleplate to floor deck, floor joists to head binder and head binder to panel rail) against the variable horizontal wind forces. The timber frame designer should, therefore, be providing a robust connection at each and every junction as part of the normal design process (Section 12.4.2).

12.6 Tying in Consequence Class 2b buildings

For timber structures in CC2b, including post-and-beam forms of construction and mass timber buildings, the provision of distinct horizontal and vertical ties is likely to be needed, and the requirements for tie forces can be achieved using Clause A.5.1 of BS EN 1991-1-7, as amended by the UK National Annex to Eurocode 1-7, Clauses NA.2.43 and NA.3.

The design tie forces appropriate to timber structures are given in Table 12.1[92].

Table 12.1: Required tying forces for robust detailing of a timber structure or a building with timber floors

Required tying forces for internal ties for robust detailing of a timber-frame building or a building with timber floors			
Horizontal design tie force, $F_{t,hor,d}$ (BS EN 1991-1-7, Clause A.5)		Vertical design tie force, $F_{t,ver,d}$ (BS EN 1991-1-7, Clause A.6)	
Internal ties (kN)	Perimeter ties (kN)	Distributed ties (N)	Concentrated ties (kN)
Max. (15 or 0.8 $(g_k + \Psi_1 q_k) s_t L$)	Max. (7.5 or 0.4 $(g_k + \Psi_1 q_k) s_t L$)	$(34A/8000) \times (h/t)^2$ or 100kN/m of wall, whichever is the greater	Each column or wall carrying vertical load should be tied continuously from the highest level. The tie should be capable of resisting a tensile force equal to the maximum design load received by the column or wall from any one storey. There should be an effective connection between vertical ties and horizontal ties at each level

g_k = dead load of floor or roof per unit area (kN/m^2)
q_k = full imposed load on floor or roof per unit area (kN/m^2)
s_t = mean spacing of ties (m)
L = full length of tied area of floor or roof (m)
Ψ_1 = frequent value of the imposed load (variable action) e.g., 0.5 for domestic floors

A = cross sectional area (mm^2) of the loadbearing wall (excluding external skin of a cavity wall) measured on plan
h = clear height of the wall (m)
t = wall width (m)
Note that the design load referred to is determined from the accidental load case

It should be noted that for both CC2a and CC2b structures, Clause NA.3.1 of the UK National Annex to BS EN 1991-1-7 allows the minimum horizontal tie force to be limited in magnitude for lightweight building structures (whose primary structure is timber or cold-formed thin gauge steel) to 15kN (Expression A.1) and 7.5kN (Expression A.2) for internal and peripheral building ties respectively, in recognition of the reduced theoretical catenary tie force from a lightweight structure over a lost supporting structural element.

In calculating the strength of connections to resist the tying forces, k_{mod} factors should be taken as the instantaneous values from Table 3.1 of BS EN 1995-1-1, i.e., $k_{mod} = 1.1$ and $\gamma_m = 1.0$ (accidental condition).

The formula for distributed vertical ties provided in Clause A.6 of BS EN 1991-1-7 was developed for masonry and concrete construction, and is onerous for timber loadbearing wall construction, which tends to be more lightweight.

For this reason, the distributed vertical tie force is rarely used in the design of timber buildings, and the notional removal of loadbearing walls is the widely-adopted approach for platform-frame buildings.

However, in the CLT industry, the provision of horizontal and vertical ties is an approach used for designing for robustness. In addition to using the more limited horizontal tie forces (as in light steel frame), the CLT industry has widely adopted the use of reduced distributed vertical tie forces, ignoring the minimum requirement of 100kN/m. Instead the vertical tie force requirement for a framed structure is used, and converted into a uniformly distributed load, distributed along the wall.

This approach is not currently supported by the codes of practice. It is intended that a working group will be set up to address the discrepancy between the current content of BS EN 1991-1-7 and the current adopted practice of the CLT industry regarding applicable tie forces.

12.7 Notional element removal

Platform timber-frame building structures which comprise loadbearing wall panels typically have a regularly-distributed arrangement of vertical walls. Notional removal of loadbearing elements (or notional panel removal), one at a time in each storey of the building, in accordance with BS EN 1991-1-7, Clause A.7, is therefore the preferred approach for checking the robustness of platform-frame construction.

In checking the robustness of timber-framed buildings, engineers need to apply judgement based on the likely three-dimensional structural behaviour of the frame, backed, where appropriate, with a two-dimensional structural assessment of discrete elements. The TF2000 full-size test building has shown this approach to be conservative, yet appropriate for platform-frame construction.

The TF2000 test building provided reassurance of the inherent robustness and availability of secondary load paths within platform frames. For example, sheathed walls with no openings can be regarded as deep beams with vertical shear taken in the panel-to-panel connections, and tension taken out through the sheathing material in continuation with any timber framework across the panel junction, e.g., via rim boards. Furthermore, the TF2000 tests demonstrated that floors have reserve strength through the transverse spanning capacity of the floor subdeck and blockings, when supported on walls parallel to the span or via upper walls acting as deep beams.

Unfortunately, the TF2000 tests were specific to one building floor and panel shape and size, so the findings cannot be used to show universal compliance with the regulations, and independent structural checks are required. It is possible to undertake such calculations to prove that wall panels can act as deep beams, but often large or numerous openings exist, leading to a requirement for additional bridging members such as rim beams.

The accidental load case should be used in design, and this is discussed in Sections 5.8 and 12.7.1.

12.7.1 Rim beam method

In this design approach, a separate engineered timber rim beam, usually installed loose on-site, is used to span between points of vertical lateral restraint (intersecting return walls or key element posts). The continuous rim beam ensures that structural continuity is achieved in the event of notional removal of a loadbearing wall panel, by providing vertical load transfer as a bridging element over the missing panel.

Using rim beams also allows joisted floor structures to be factory-assembled as cassettes, with a rim board used to connect the joist ends together for transportation. The rim beam can later function as a vertical load transfer element in the completed structure. An example of the design checks for the rim beam structural methodology is set out in Section 12.7.2.

Generic guidance can be found in '*Verification of the robustness of a six-storey timber-frame building*'[89].

The rim beams need to react onto wall intersections and the wall returns must be of 1,200mm minimum length (excluding framed openings). Such walls can be non-loadbearing in the conventional sense, but must still be capable of transferring loads down through the structure; the use of lightweight partitions built off floating floors is not acceptable. Any rim beams used are supported at the wall intersections by corner stud groups. For the studs to provide proper support, the beams must bear fully onto them. To achieve this, the wall panels should be lapped in the opposite manner to the rim beams. If no stud clusters are present below the rim beam bearing, hangers or proprietary disproportionate collapse brackets need to be provided off adjacent rim beams, to ensure proper support at the wall intersection.

Figures 12.2–12.6 indicate typical arrangements of rim beams and supports to achieve robustness.

Figure 12.2: Typical plan rim beam layout

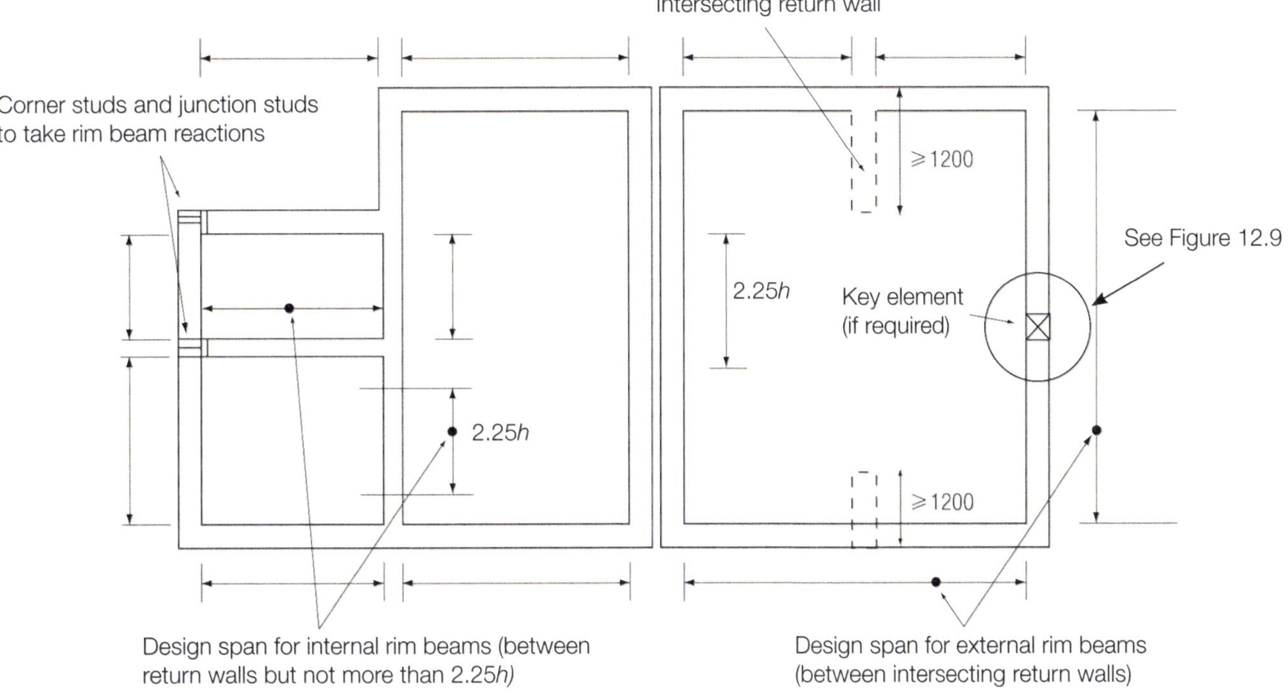

Figure 12.3: Plan view of a typical rim beam layout indicating recommended arrangement of rim beam laps at wall intersections

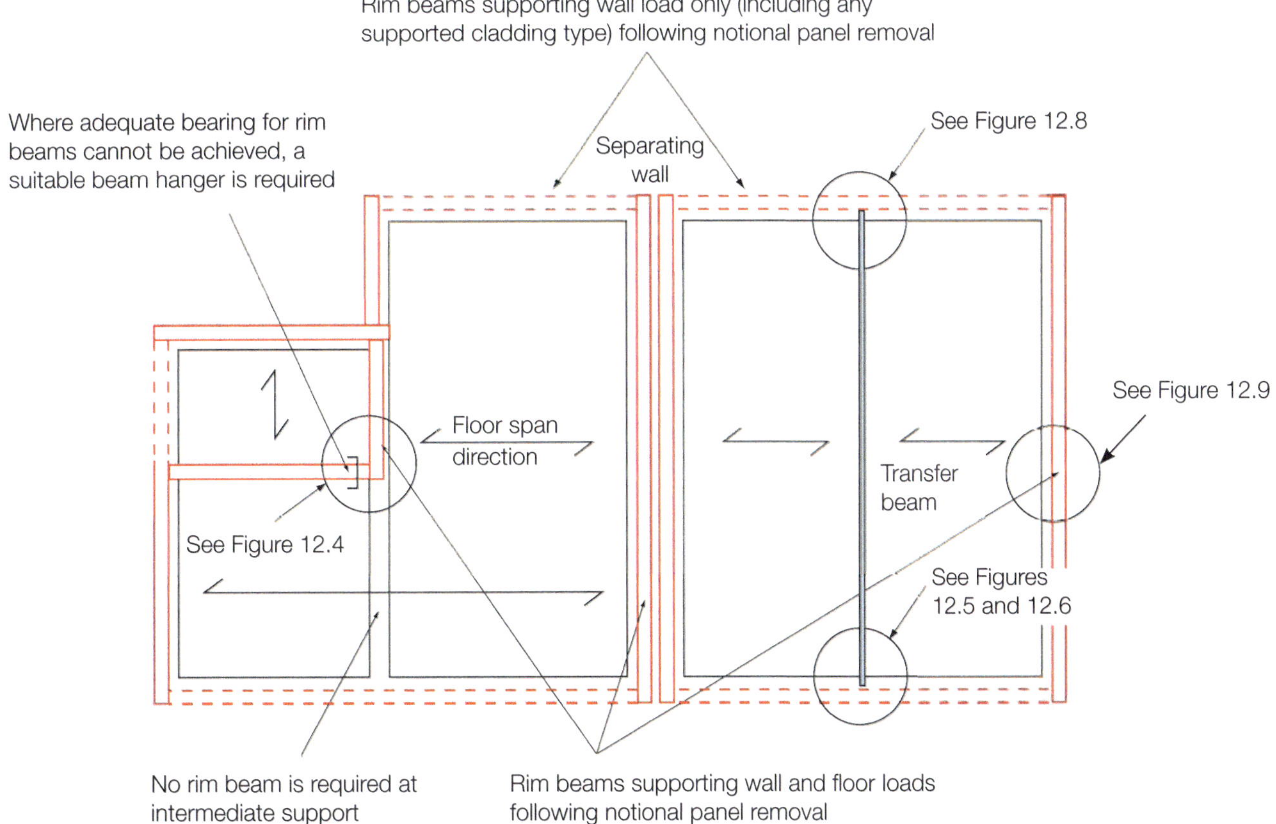

Figure 12.4: Rim beam junction at a corner intersection

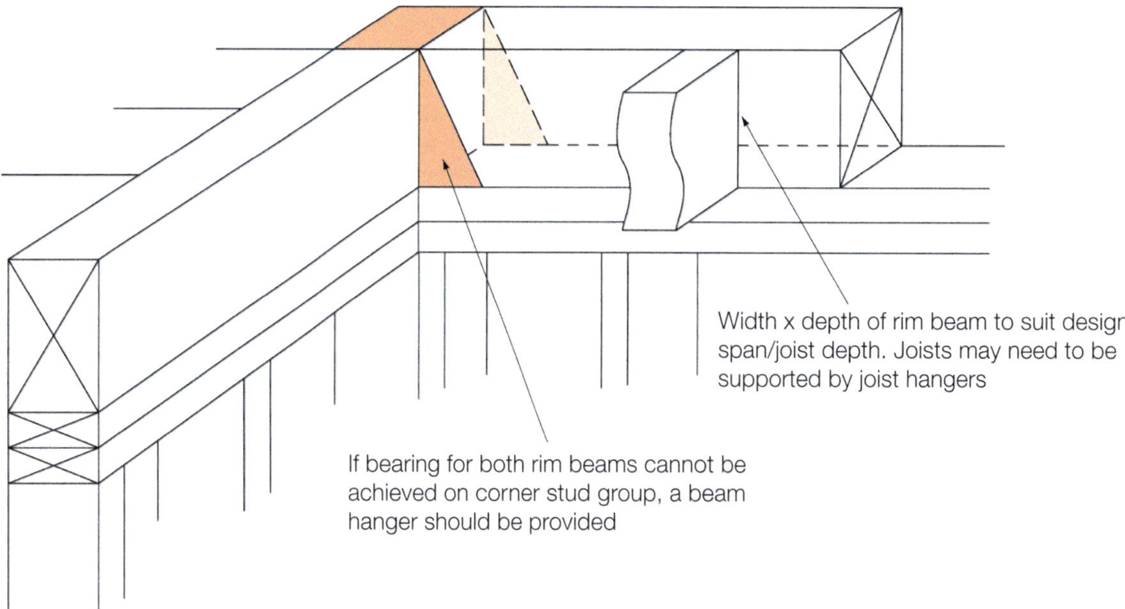

Figure 12.5: Transfer beam/rim beam junction Option 1 — no key element post

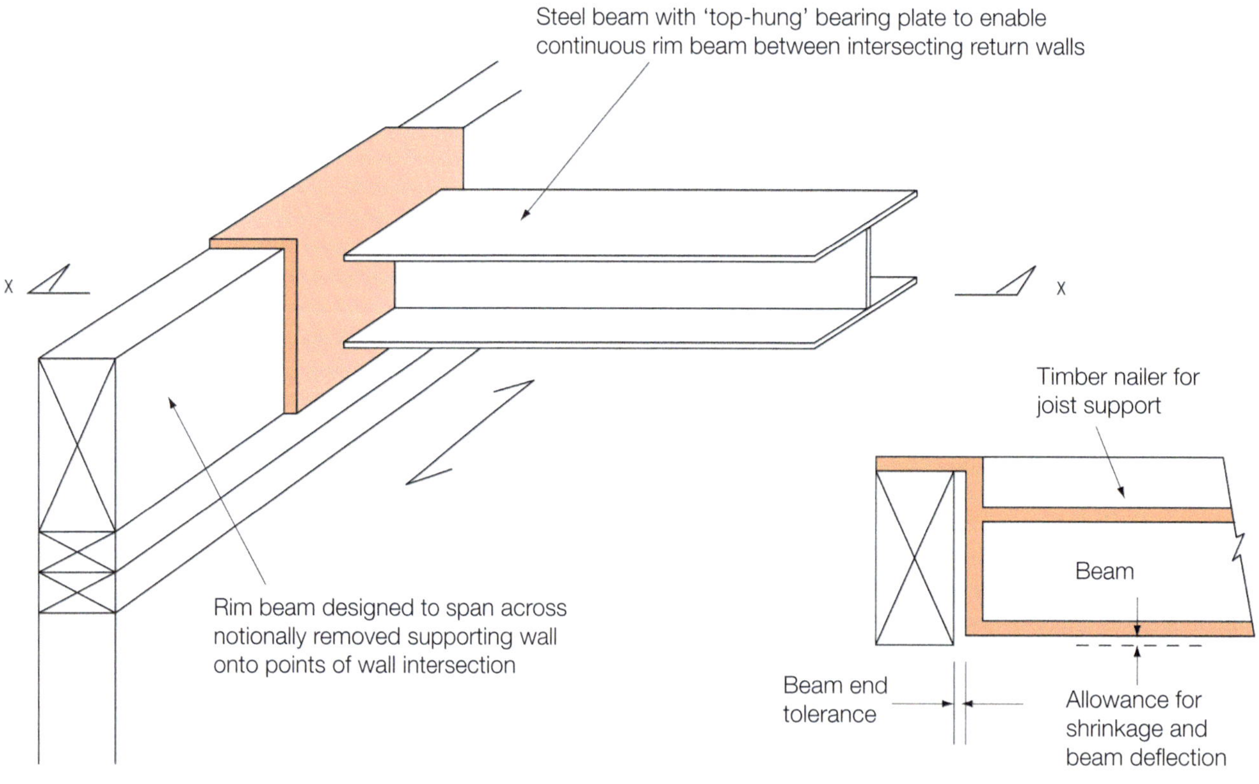

Figure 12.6: Transfer beams within timber structures

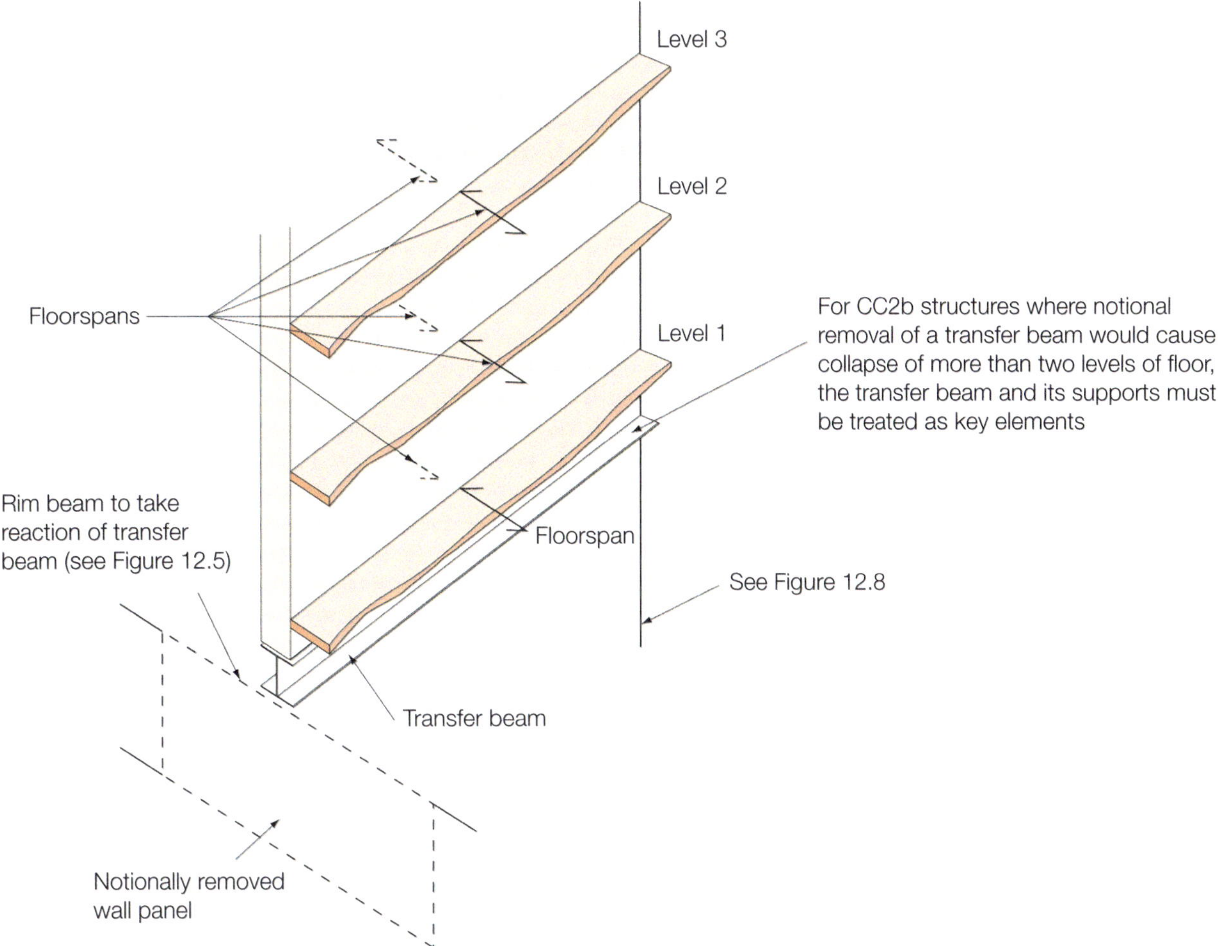

If the structure contains transfer beams, and the notional removal of a transfer beam would cause unacceptable collapse, the transfer beam will need to be designed as a key element. Where the rim beams cannot be designed to span the required distance between return walls, one or more intermediate posts will be needed and treated as a key element and designed accordingly (Section 12.8).

For design load cases appropriate for the design of both rim beams and key element posts, BS EN 1990[34], Clause A1.3.2 and Table A.1.3, and the supporting UK National Annex[44], Clause NA.2.2.5 give guidance on accidental load combinations and psi (Ψ) values to be used for the design of elements subject to accidental design situations (Section 5.8 of this *Guidance*).

In a residential building, these would be:

Floors
- 1.0 permanent + 0.5 imposed
- 1.0 permanent + 0.2 wind + 0.3 imposed

Roof
- 1.0 permanent + 0.2 imposed
- 1.0 permanent + 0.2 wind

In addition, BS EN 1995-1-1, Table 3.1 gives k_{mod} for timber-based materials applicable to accidental actions. Values for short-term actions are appropriate for the residual loading effects following an accidental action, as described in Table NA.1 of the UK National Annex to BS EN 1995-1-1.

12.7.2 Design checks for the notional element removal approach

The following structural checks should be undertaken when designing a rim beam to span over the notionally removed element — the wall panel — and assessing the impact of the removal of an internal wall, by calculating the capacity of the joists to span over the removed wall.

1. The nominal length of wall should be determined.
   For an external wall, the nominal length of wall is the clear length between lateral supports (intersecting return walls), or members designed as key elements.

   For an internal wall, the nominal length of wall is a length not exceeding 2.25h where h is the clear height of the panel between lateral supports (the top of the structural deck level below to the underside of the structural joist level above (Figure 12.7).

2. The impact of the removal of a wall providing joist support should be checked.
   Firstly the collapse area should be assessed. Unless the joists are top-hung from the rim beam, the joists immediately above the removed wall panel are assumed to collapse, and the resulting collapse area needs to be less than 15% of the total floor area or 100m².

   Secondly, the capacity of the rim beam to support a single-storey of wall (plus any supported cladding) and the floor joists at the level above, to span over the notionally removed wall panel should be checked (Fig. 12.7). Subsequent rim beams provide support to other walls and floors, preventing propagation of the collapse.

3. The impact of the removal of an internal wall should be checked.
   Following the notional removal of an intermediate loadbearing wall, continuous floor joists (A–B in Fig. 12.7) act as a double span at each level. The joists need to be checked for the design residual floor loads and the self-weight of a single-storey of wall (C–D in Fig. 12.7).

Figure 12.7: Notional panel removal

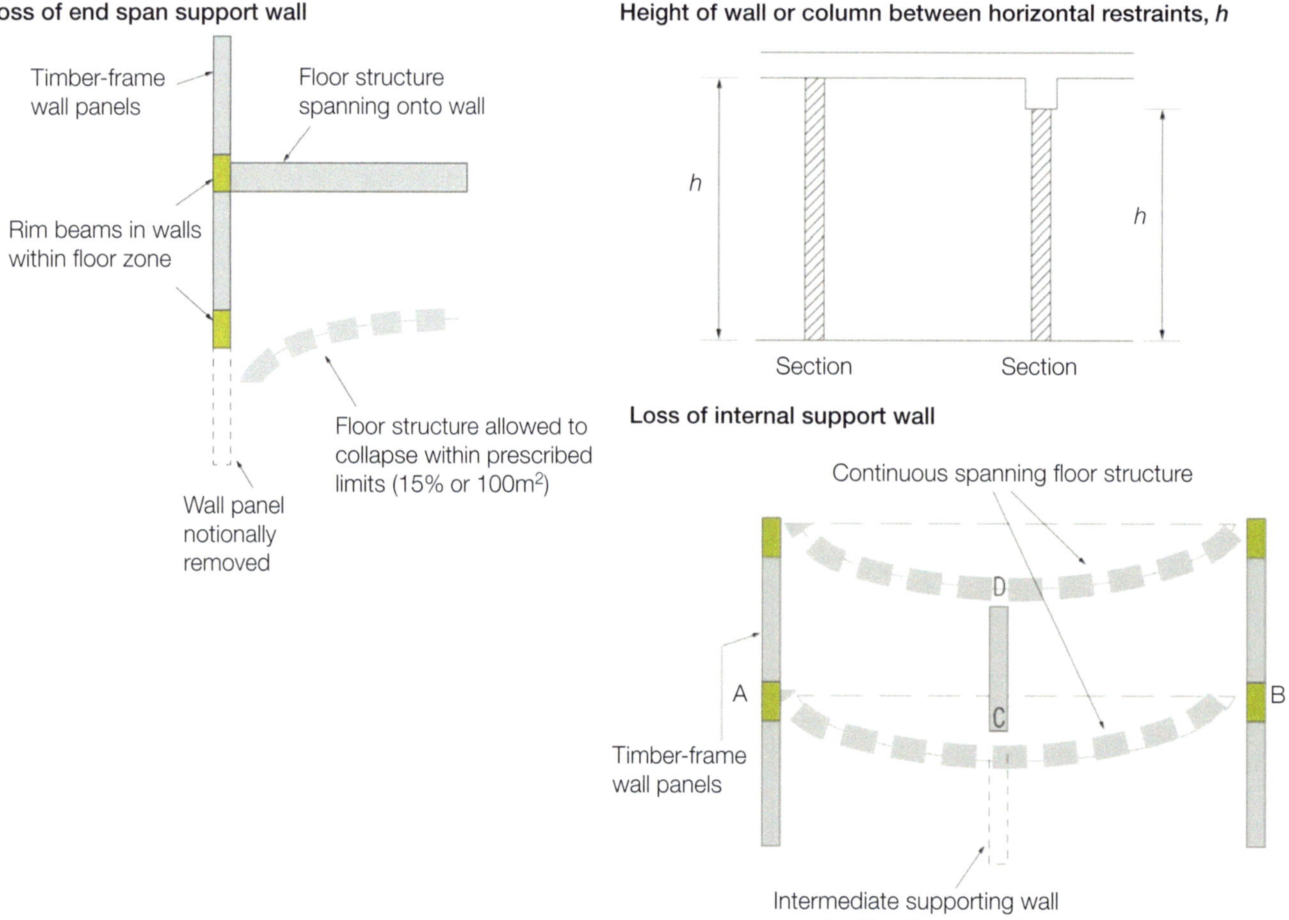

12.8 Key element design

Key element posts (KEPs) are often provided in combination with rim beams, either to reduce the length of wall panel that must be considered for notional panel removal, or to support transfer beams within the timber-frame structure (Figure 12.8).

Figure 12.8: Transfer beam/rim beam junction Option 2 — with key element post

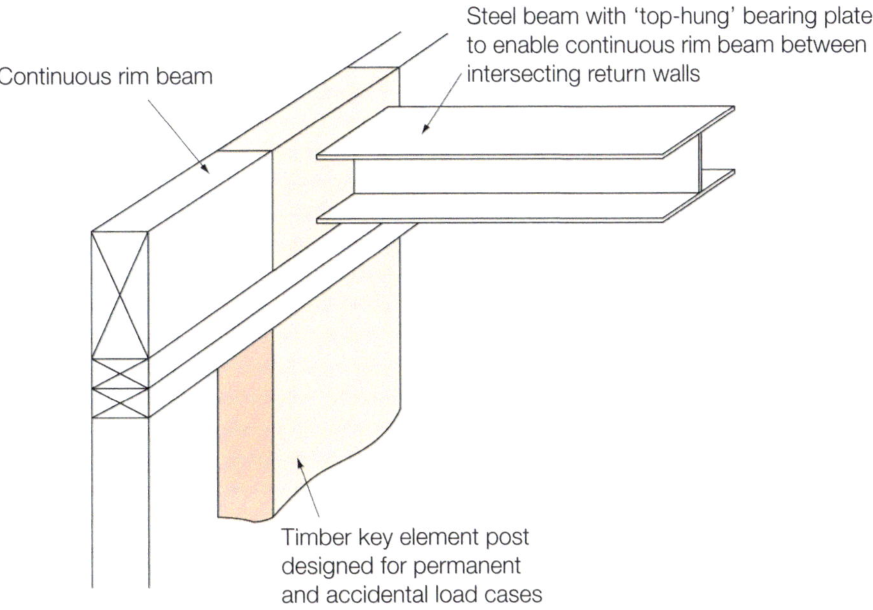

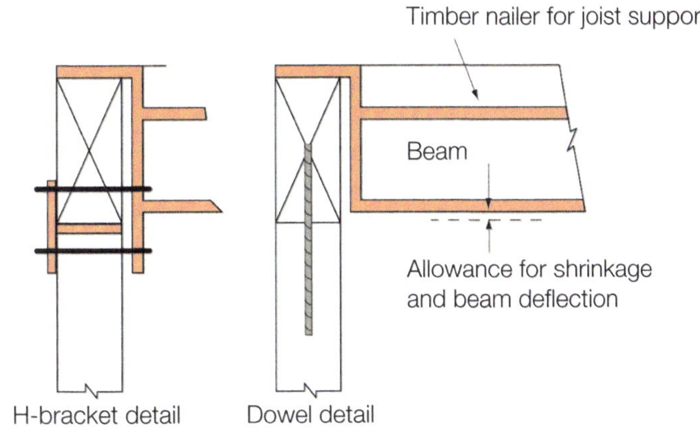

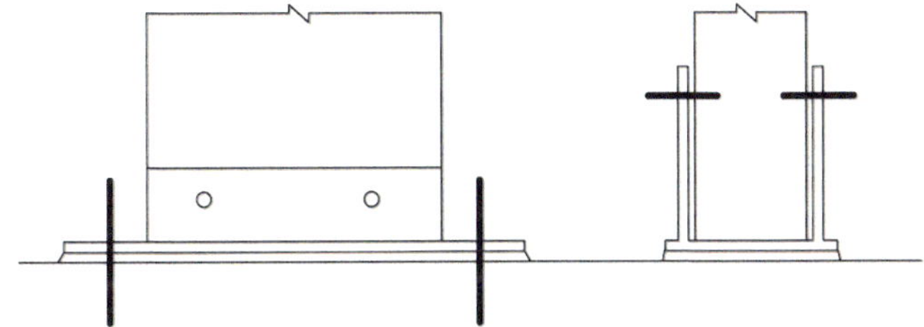

KEPs are typically provided as engineered timber, e.g., glulam or laminated veneer lumber (LVL) materials within a timber-frame wall panel, with brackets connecting the KEP to the rim beams at each floor level, to transfer the accidental loads back into the floor diaphragm.

KEPs are required to be designed for an accidental load of 34kN/m^2 (BS EN 1991-1-7, Clause 3.3) applied over the width of the post in one direction at a time. The additional reaction of adjacent wall panels, which are connected to the KEPs with nails, is likely to be small, but should be considered when checking the capacity and restraint at floor levels (Figure 12.9).

Figure 12.9: Key element posts acting in conjunction with rim beams

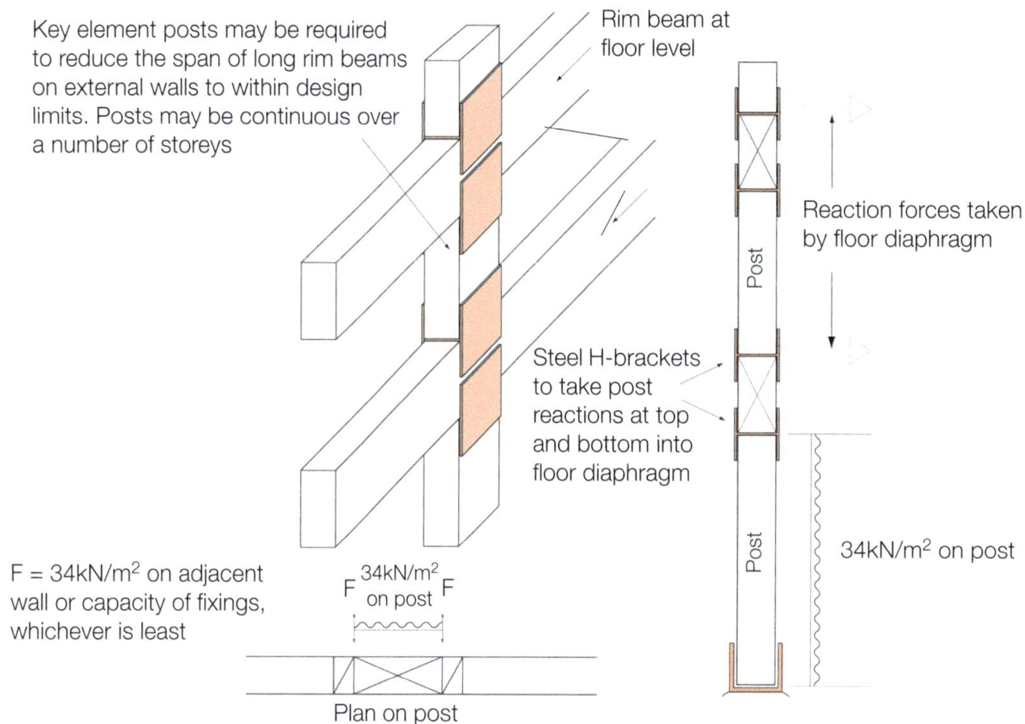

Transfer beams occur in timber structures supporting large floor areas. For general guidance on the design of these elements, reference should be made to Section 5.7.1 of this *Guidance*. Prudence suggests checking these as key elements, and this can be done by checking supporting members and providing robust connections at top and bottom (capable of resisting the stipulated 34kN/m^2). Fig. 12.9 indicates typical arrangements of KEPs and beams within loadbearing wall structures.

12.9 Robustness of typical structural forms

The approach for designing for robustness will depend on the structural form and material of the building, whether solid timber, engineered timber, mass timber, etc. This section sets out the design considerations for different types of timber buildings.

12.9.1 Trussed rafters

Tests by BM TRADA in 1993 undertaken on a 10m span roof comprising 25° fink trusses indicated that trussed rafter roofs have inherent robustness[93]. This was found to be the case both for suspended and plasterboard ceiling arrangements, and under unfavourable arrangements of wall plate end joints occurring above removed supporting walls. The four secondary components found to contribute most substantially to resisting collapse following wall removal were the wall plate, chevron bracing, top chord nodal bracing and even the tile battens with their interlocking tiling. Although the tests were of trussed rafters, it is considered that bolted trusses would behave in a similar manner.

Experience from the US[94] indicates that most failures in timber trusses occur in individual members, as a result of member shrinkage and connector slip, yet do not result in the failure or collapse of the truss structure overall. Usually there is some readjustment among other framing members, so that stresses are redistributed through the bracing members and, therefore, bracing plays an important role.

Part A3 requirements make no direct reference to roofs but the now withdrawn British Standards for steelwork (BS 5950-1[63]) and masonry (BS 5628-1[83]) both incorporated a clause permitting the exclusion of horizontal ties in roofs of lightweight construction. Therefore, it may be argued that collapse of light timber roofs onto the floor below is unlikely to lead to floor failure and progressive collapse. However, in all timber construction, the ability of floors to sustain falling trusses may be less likely than in other materials, and for this reason the design of timber floors below roofs should consider the possibilities of debris loading (Section 5.6.2). Because timber elements are so light, they are more easily dislodged (e.g., roof trusses under wind suction), so fixings and restraints are especially important for assuring robustness, and caution is required in assessing the consequences of a roof failure.

12.9.2 Platform timber-frame construction

Structures comprising loadbearing walls such as platform timber frames typically use materials easily fastened together, and the benefits for robustness have been demonstrated by full-scale testing[90]. However, despite the potential contribution of secondary items, such as claddings and linings, to the overall robustness of a structure, engineers should be relying on properly-engineered load paths, and not on potential load paths through finishes which may or may not be there depending on how the secondary items are fixed. Reliance on connections is only justified if the connection detailing rules of the relevant codes of practice are complied with.

Platform timber-frame construction relies on a cellular plan form with all wall and floor components fixed to each other. Reliance is placed on the diaphragm action of the floors to transfer horizontal forces to a distributed layout of loadbearing walls. In turn, these walls combine vertical support with horizontal racking resistance. The format is inherently robust, proved by many years of experience and through full-scale tests. Robustness of the structural form exists as long as the following general principles are adopted:

- Provision of structural units that can be tied together. In particular, the fixing together of all intersecting wall panels should be designed in accordance with Clause 20.5.2.9 of PD 6693-1. Previously, Clause 4.9.2 in the now withdrawn BS 5268-6.1[95] stated a minimum fixing density for wall panels of 3.35mm diameter nails of length 75mm at 300mm centres. This should be considered the minimum fixing density for wall panel connections
- Ensuring that layouts and plan arrangements provide returns and intersecting walls and floors. Return walls should be provided at the ends of all loadbearing walls

Consideration should be given to the layout of loadbearing walls early in the design process, to assist with providing a robust structure. Cellular layouts are best suited to multi-storey platform timber frames. Internal loadbearing walls may need to be strengthened to carry horizontal forces and, where party walls separate the structure into separate units, the design must ensure that horizontal forces can be taken by each unit. The opportunities for load transfer across party walls via structural ties are limited, for reasons of maintaining acoustic performance as discussed in BRE Report BR454.

Open-plan layouts with no transverse structure are inappropriate for platform timber frames, and additional structure is required for stability. The introduction of portal frame elements can provide solutions to open-plan layouts, but consideration should be given to appropriate deflection limits and connectivity of the framing types, as well as controlling differential movement of different materials.

The timber codes Eurocode 5 and PD 6693-1 provide appropriate guidance for achieving strength and stiffness of the components that make up multi-storey timber frames, and for the provision of overall building stability, and meeting these requirements is a pre-requisite for robustness. In particular, racking resistance, overturning and resistance to sliding should be checked.

In addition to designing a structure for stability by considering the equilibrium of forces and the strength and stiffness of elements, the robustness of a building is also achieved through good practice detailing, to ensure connectivity of items providing additional stiffness to the building. This is described in more detail in Sections 12.4 and 12.5.

Consideration must also be given to the asymmetric layout of elements intended to provide resistance to horizontal forces (e.g., racking walls). For structures or layouts that are unable to provide sufficient racking resistance, or have an unbalanced arrangement of racking resistance on-plan, the use of discrete stiffening elements such as rigid frames, should be considered. The deflection limits for rigid frames should be appropriate for the structure and applied finishes. A limit of at least height/300 or height/500 where finishes sensitive to movement are supported by the frame should be adopted.

The TF2000 tests demonstrated that the actual stiffness of a timber-frame building is significantly higher than the minimum values demanded by the code. Based on TF2000, there is evidence that current design principles are adequate for strength and stiffness for buildings up to eight storeys, and that there is no reason for normal cellular platform frames to have additional deflection limits imposed.

Once the engineer is satisfied that the building has a robust and stable layout, a decision on how the building will satisfy the disproportionate collapse rules should be made. The usual approach for platform timber-frame buildings is to take the notional element removal approach.

12.9.3 Mass timber construction (e.g., cross-laminated timber)
As with other forms of construction, the approach to demonstrate the robustness of mass timber buildings will depend on the structural form and constraints of the building. A common route to compliance is to design the structural elements to resist the prescriptive rules for horizontal and vertical tie forces, as outlined in BS EN 1991-1-7. However, there is a discrepancy between the current content of BS EN 1991-1-7 and the current adopted practice of the CLT industry regarding applicable tie forces. The STA intends to set up a working group to address this.

12.9.3.1 Recommended tie forces for mass timber construction
While BS EN 1991-1-7, Clause A.5.2 provides equations to derive tie forces for loadbearing forms of construction, these equations have been conceived with heavy forms of construction in mind (specifically steel and concrete) and these forces are not considered appropriate to CLT loadbearing wall construction.

The equations provided in BS EN 1991-1-7, Clause A.5.1 to determine the horizontal and vertical tie forces for framed structures as indicated in Table 12.1, are based on the permanent and variable actions on the structure, but also include default minimum values. Since concrete and steel structures are normally heavier than an equivalent timber structure, the minimum forces provided are likely to be the governing tie forces to be used for design, and can be onerous for timber structures.

For mass timber buildings using framed structural arrangements (e.g., engineered timber post and beam type structures) the horizontal tie forces provided in Expressions A.1 and A.2 can be reduced, according to the National Annex to BS EN 1991-1-7, Clause NA.3.1, which provides a more achievable means of satisfying the tie force requirements (Section 12.6).

For mass timber loadbearing wall structures, the accidental tie forces typically used are as given here and in Figure 12.10:

- Distributed internal ties: $T_i = \text{Max. } (15, 0.8(g_k + \Psi q_k) \times L)\ (\text{kN/m})$
- Distributed perimeter ties: $T_p = \text{Max. } (7.5, 0.4(g_k + \Psi q_k) \times L)\ (\text{kN/m})$
- Distributed vertical ties: $T_v = g_k + \Psi q_k\ (\text{kN/m})$

CROSS Safety Report: The risk of collapse of multi-storey CLT buildings during a fire[96]
This report highlights a growing trend in the industry regarding fire safety of buildings comprised of cross-laminated timber (CLT) structures. The concern particularly relates to multi-storey sleeping risk buildings in the UK.

Figure 12.10: Tie forces for CC2b structures of mass timber loadbearing wall construction (where notional panel removal is not appropriate)

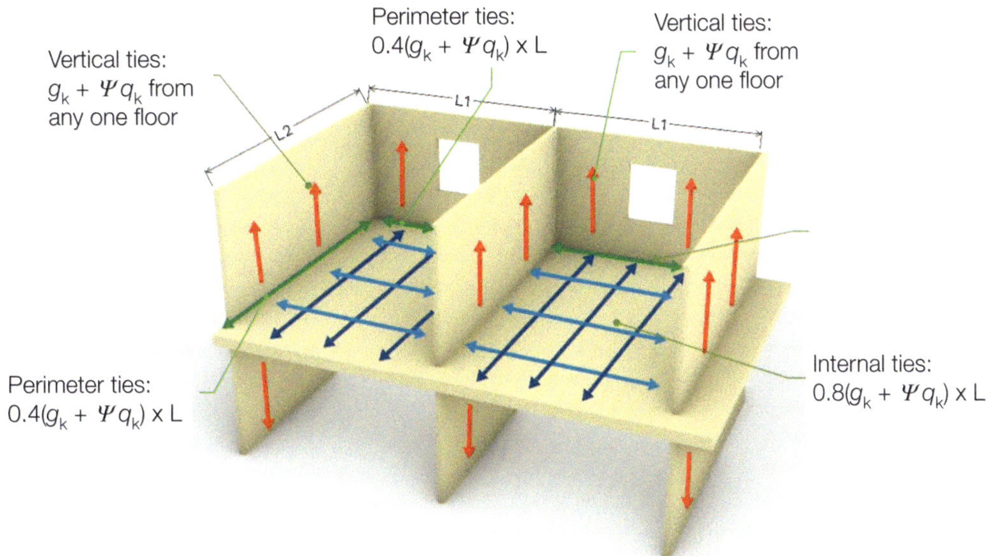

12.9.3.2 Provision of horizontal and vertical ties in mass timber structures

There is some debate about the origins of the equations that state the minimum tie forces for design, and whether they are based on catenary action (Figure 12.11). Following the loss of a supporting member, the connection of a beam or floor slab would require significant ductility and continuity of load paths to allow catenary action to function appropriately.

Figure 12.11: Development of catenary tie forces following an unspecified accidental event

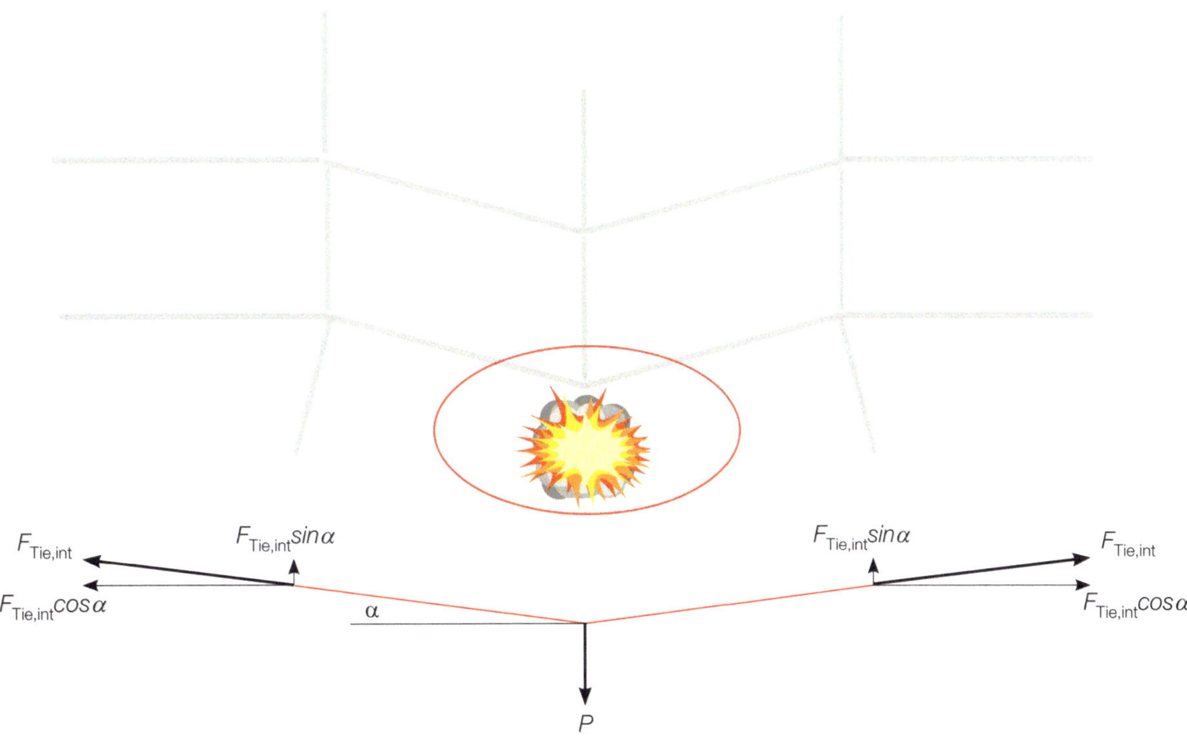

The Eurocodes do not provide any guidance on the minimum ductility requirements for connections, or how continuity of load paths can be maintained when using the provided tie forces, but it is recommended that the engineer considers these effects when demonstrating robustness by this route.

12.9.3.3 Connection design in mass timber structures

The engineer should be aware that the robustness requirements will often be the governing load case for the design of connections. As part of the outline design of the building, it is essential that a clear understanding of how robustness is to be achieved is developed so that appropriate forces can be identified for connection design. Unlike in steel or concrete design, it is most unlikely that a timber design will satisfy robustness requirements by default.

Vertical ties should be continuous from foundation to highest level, including connection to the substructure. This timber-concrete connection will often prove to be onerous and should be considered early in the design process. It will often be necessary to specify tension-resisting brackets, the dimensions of which may affect finishes or require special machining to the wall panels.

12.9.4 Framed structures — glulam or engineered timber post-and-beam and traditional carpentry frames

Open-plan structures may be constructed as post-and-beam timber-frame structures. Provision of stability in the two orthogonal directions, temporary stability during construction and adequate loading allowances should be sufficient to ensure buildings are robust in both permanent and temporary conditions.

For buildings where the failure of a single member (e.g., a column or principal truss) could cause a catastrophic collapse out of proportion to the element size, a relatively high degree of robustness should be provided. This can be achieved by designing the connections for the horizontal tie forces required in Annex A.5 of BS EN 1991-1-7.

For framed structures the following accidental tie forces as described in Table 12.1 are appropriate:

- Concentrated internal ties: $T_i = \text{Max. } (15,\ 0.8(g_k + \varPsi q_k) \times s \times L)$ (kN)
- Concentrated perimeter ties: $T_p = \text{Max. } (7.5,\ 0.4(g_k + \varPsi q_k) \times s \times L)$ (kN)
- Concentrated vertical ties: $T_v = g_k + \varPsi q_k$ (kN)

In addition, timber columns should be provided with fixed bases that are able to carry a moment and shear force equivalent to that resulting from the notional horizontal loads (Section 12.4).

Single-storey post-and-beam frames should have roof structures adequately tied to the supporting structures, in accordance with the guidance given in Section 12.5.

Where mechanical fasteners are already provided to form member connections, the tie forces can generally be accommodated by existing connections, designed in accordance with Eurocode 5, as the partial safety factors for duration of loading and material strength for accidental actions typically make this load case less onerous than for wind and vertical load.

A special form of post-and-beam frame is traditional carpentry frames, using pegged mortice and tenon joints. As most of these structures will fall into CC1 they will require no special considerations for robustness. However, where they are required to satisfy the requirements for robustness in a higher structural consequence class, it might be necessary to consider replacing joints with mechanical fasteners that can be designed to resist tie forces, or applying a risk-based approach to robustness similar to the provisions for CC3.

12.9.5 Timber portal frames

As for steel buildings, a special case arises for large-span portals, since failure of a single frame can result in a considerable collapsed area, potentially putting the structures into CC2a or CC2b.

There are similarities and differences between timber and steel portal frames. Steel portals have joints capable of plastic action, which will differ from timber frame behaviour which should be assumed to behave elastically. The eaves joint on a timber portal frame must also be checked for combined bending and compression, to ensure that lateral torsional instability of the timber sections does not occur. This is similar to the checks that would be undertaken for steel portal frames, and is especially important where deep, slender sections are proposed. This design check should be carried out in accordance with Clause 6.3.3 of BS EN 1995-1-1.

Robustness should also be checked by assuring that there is no risk from longitudinal instability (e.g., by failure of wall bracing systems) and that the horizontal restraint to the portal base is adequate to resist both the forces from the permanent design load cases and the notional horizontal load case (Section 12.4).

12.10 Timber and fire robustness

Fire robustness is discussed in Chapter 14, but because timber is a combustible material it may need further consideration as it can be consumed during a fire, and also contribute fuel to a fire. Timber structures and structural products are also vulnerable to fire spread over their surface.

Structural robustness in fire depends on the severity of the fire event and consequence of failure. If the consequence of building failure is great, a high level of robustness is needed. For low-rise and low-consequence buildings, the collapse of a structural member or part of a building due to fire may be acceptable. For high-rise and high-consequence structures, or buildings with vulnerable residents, any collapse may be unacceptable or need to be delayed so that escape is possible.

The amount of timber in a building can directly impact the behaviour and severity of a fire. For buildings constructed with little timber, a fire event is not likely to be greatly affected by the presence of the timber, and the fire is influenced more by room contents and ventilation. In a building with a large amount of timber, whether structural or not, the timber can contribute fuel to the fire, unless it is protected in a way that means the fire cannot spread to the timber.

As a timber structural member burns, there is a loss of section which results in a reduction in stiffness and strength within the remaining structure. Where it is critical that a structural timber member survives the duration of the fire to prevent building collapse, timber sections may need to be increased in size.

Most connections in timber are made with metal components such as nails, screws and bolts, sometimes used in conjunction with steel plates. Steel quickly loses its strength at elevated temperatures, so steel connections between timber elements, even if the steel is only partially exposed, are highly vulnerable in a fire.

In CLT design, a recurring theme is the issue of a floor needing to act as a diaphragm during fire conditions. Typically, in hybrid steel and CLT construction, the screws transferring loads between the CLT and the steel frame are lost during the fire, meaning that the fire-affected floor cannot act as a diaphragm. If the columns cannot be justified to span vertically over two floors, then an alternative means of securing the CLT to the beams is needed in the fire case.

Loss of fasteners due to fire can also affect other horizontal or vertical bracings and have an impact on the stability of all timber buildings during a fire.

Guidance on the use of BS EN 1995-1-2 and updates on state-of-the-art knowledge is provided on the *Fire Safety Wood in Construction* website[97].

 *CROSS Safety Report: Cross-laminated timber (CLT) in multi-storey buildings*[98]
This article discusses use of CLT in buildings and how its use cannot be considered a common building situation, and that it can contribute fuel to a fire.

 *We need to talk about timber: fire safety design in tall buildings*[99]

 *Burnout means burnout*[100]

12.10.1 External wall construction and statute
In 2018, Regulation 7 of the Building Regulations was amended to ban combustible materials from being used in external walls for buildings with an upper storey over 18m and with specified uses, including residential. These banned materials include timber.

The guidance of Approved Document B[101], which is not mandatory but typically used for 'common building situations' to demonstrate compliance with the Building Regulations, initially limits the height of structural timber external walls to an upper storey height of 11m, unless evidence can be provided to show that the external wall structure is fire-safe between 11m and 18m upper storey heights.

A common building situation is not defined, but is likely to be where there is long experience of the building material type, and fire behaviour is known. Mass timber construction (e.g., CLT) is not considered to be a common building situation for complex buildings and those above the 11m floor height threshold.

Therefore, for CLT to be used in external walls, evidence for fire safety will need to be checked by completing a performance-based design justification. The outcome of that design check may be the conclusion that CLT is not suitable for buildings with a floor level above 11m.

12.11 Alternative methods of achieving compliance for timber structures

The Eurocodes are currently being updated and with more coherent guidance on achieving robustness of building structures should be in use no later than 2028. While the provision of horizontal and vertical ties designed to prescriptive tie forces will still be an acceptable approach for demonstrating robustness, the recommended options for timber structures will be to adopt notional removal of elements or to design key elements. Engineers will still be able to use tie forces as a route to compliance but, as a consequence, would be required to consider the ductility of connections and continuity of load paths as a result.

It is therefore recommended that designers give consideration to demonstrating robustness using notional element removal and, where this is not possible, to design key elements, similar to the methods described in Sections 12.7 and 12.8. Using this approach would require the engineer to arrange the structural elements to be supported across multiple spans.

Timber is a lightweight form of construction (even mass timber) where the imposed load is commonly a high proportion of the total load, and serviceability considerations are usually the governing design criteria. Under accidental loading conditions, where serviceability considerations are less onerous, it might be possible to demonstrate that a floor structure or beam can cantilever over an internal support or between two outer supports (Figure 12.12). However, this may mean that floor structures cannot be broken at party wall lines in residential buildings, which may affect the choice of acoustic finishes.

Where CLT floor elements are used, it is frequently not possible to introduce a rim beam that can span perpendicular to the primary span direction. While CLT can span in two directions it is unlikely to be able to span far enough to offer architects sufficient freedom in their floor plan. A better approach is to fix the floor to the wall elements above, and design the wall to act as a deep beam between return walls. This is straightforward where the wall panel is made from a single full-width structural element, but becomes more challenging if the wall is made from many smaller structural elements or contains openings. In this scenario the engineer should design the multiple wall elements to act compositely, as a mechanically-jointed diaphragm, to span over the notionally removed opening between return wall elements.

Other possible methods of achieving support for walls and floors following the notional removal of wall panels are:

- The use of room-size floor 'cassettes' with cassette edge boards acting as rim beams, bridging over removed wall panels. This is possible where small, repeatable room sizes are present (e.g., hotel type accommodation)
- Loose-floor construction with top-hung floor joists or joists supported in joist hangers from loose rim beams which bridge over removed wall panels
- Where joist span lengths are repetitive, using alternate double-spanning joists with selected walls designed as deep beams. A check should be made that overall building instability does not occur due to lack of restraint at the back span of any cantilevered joists
- Cantilevered joists or CLT slabs supporting walls where the floor structure is designed to support the point load reaction from a single storey of wall panel, plus any supported claddings following notional panel removal. Fig. 12.12 shows the structural methodology applicable to the cantilevered floor approach for satisfying disproportionate collapse

Figure 12.12: Disproportionate collapse philosophy — cantilevered floor method (notional removal of an external or compartment wall)

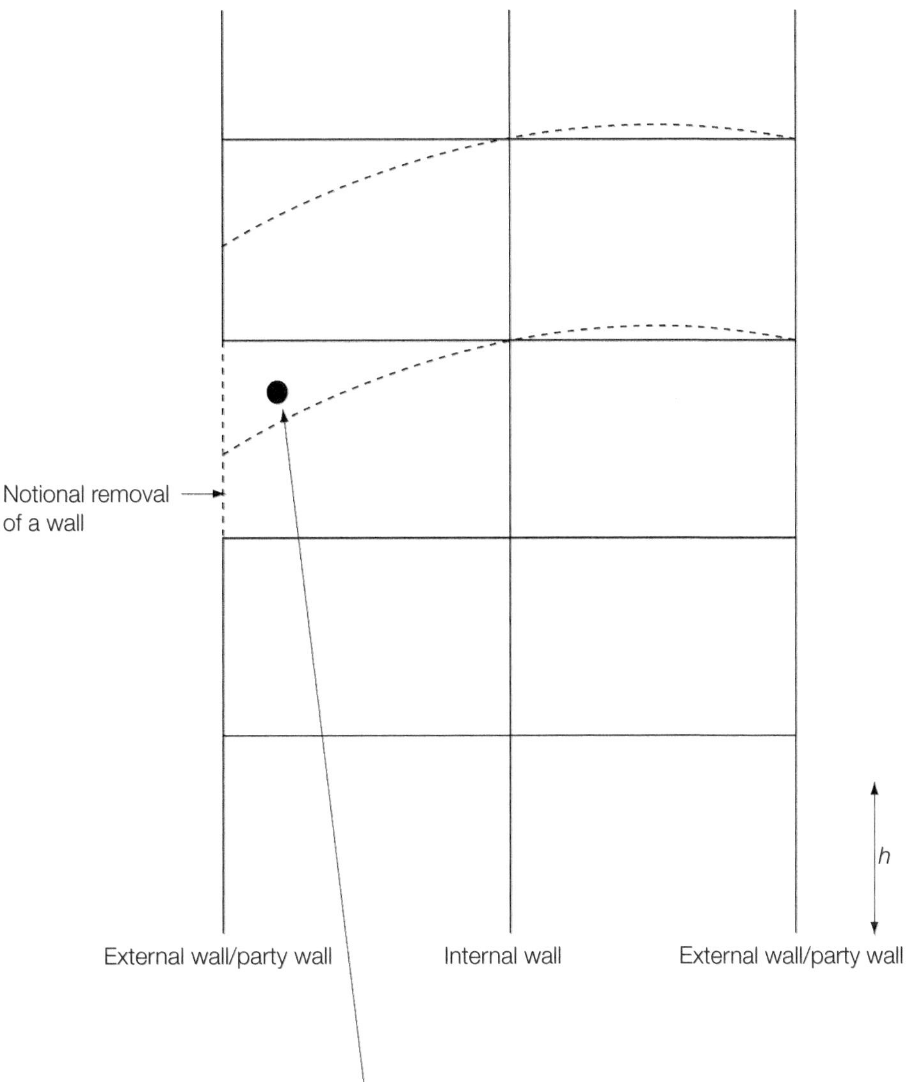

Following notional removal of a wall providing joist end-span support, the joists are designed to cantilever and support a storey height of wall panel at the cantilever tip. Sufficient holding-down resistance is required to prevent back-span uplift (which can be reduced by staggering joist layouts)

Note that for the case of the removal of an internal loadbearing wall, the accidental load case for the joists is the same as for the rim beam method, with joists designed for an increased span length supporting the weight of the non-loadbearing partition at mid-span.

Case study 13

Notional panel removal and engineered timber rim beam design

This case study looks at notional panel removal and engineered timber rim beam design according to BS EN 1995-1-1 and BS EN 1991-1-7. The rim beam within the floor zone is located at the end of a 5.0m span floor cassette. The floor structure has a self-weight of 1.00kN/m^2 and a design imposed floor load of 1.50kN/m^2. The rim beam is located within an external wall panel which has a storey height of 2.75m and a self-weight (including supported cladding) of 0.50kN/m^2.

Following the notional removal of a 4.5m length of the external wall panel between intersecting return walls, check that the rim beam will meet the ULS requirements of BS EN 1990 and BS EN 1995-1-1 for the residual structure following an accidental event. The rim beam is assumed to be prevented from lateral buckling by the connections of the wall panels and soleplates above the beam.

Geometry of structure (Figure 12.13)

Figure 12.13: Structural details

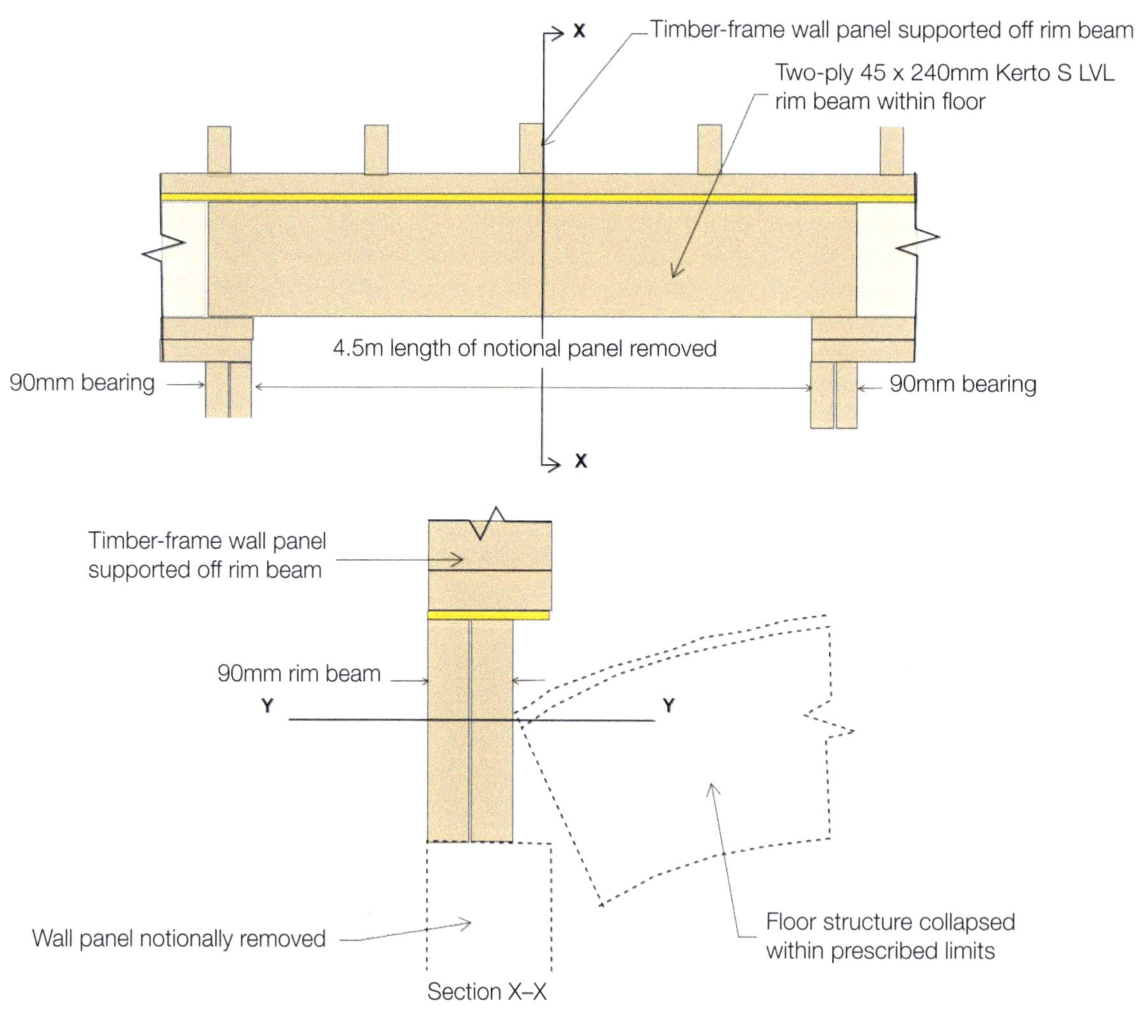

Beam geometric properties

Breadth of the rim beam =

$b = 2 \times 45\text{mm} = 90\text{mm}$

Depth of the rim beam =

$h = 240\text{mm}$

Clear span of the rim beam =

$l_c = 4{,}500\text{mm}$

Bearing length of the rim beam at each end onto C16 timber =

$l_b = 90\text{mm}$

Design span of the rim beam =

$l = l_c + l_b = 4{,}590\text{mm}$

Section modulus of the beam about the Y–Y axis =

$w_y = \dfrac{(b \times h^2)}{6} = 8.64 \times 10^5 \text{mm}^3$

Beam material properties
(From Metsawood Kerto S LVL VTT Certificate No. 184/03 March 2009)

Characteristic bending strength of LVL =

$f_{m,k} = 44\text{N/mm}^2$

Characteristic shear strength of LVL =

$f_{v,k} = 4.1\text{N/mm}^2$

Characteristic bearing strength of LVL =

$f_{c,90,k\,(LVL)} = 6.0\text{N/mm}^2$

5th percentile modulus of elasticity parallel to grain =

$E_{0.05} = 13{,}800\text{N/mm}^2$

Mean shear modulus =

$G_{0,mean} = 400\text{N/mm}^2$

Mean density =
(refer to BS EN 338[102] Table 1 for C16 strength classes)

$\rho_m = 510\text{kg/m}^3$

Characteristic bearing strength of C16 =

$f_{c,90,k\,(C16)} = 2.2\text{N/mm}^2$

Partial safety factors

Permanent actions =
(UK National Annex to BS EN 1990 Table NA.A1.3 for accidental actions)

$\gamma_{G,ULS} = 1.0$

Factor for the frequent value of the lead variable action =
(UK National Annex to BS EN 1990 Table NA.A1.1
for Category A variable imposed actions)

$\Psi_1 = 0.5$

Material partial safety factor for LVL and C16 =
(UK National Annex to BS EN 1995-1-1 Table NA.3)

$\gamma_m = 1.0$

Actions

Gravitational acceleration =

$$g = 9.81 \text{m/s}^2$$

Characteristic self-weight of the beam =

$$G_{k,swt} = g \times b \times h \times \rho_m = 0.11 \text{kN/m}$$

Wall self-weight is taken as

$$g_{k,wall} = 0.50 \text{kN/m}^2$$

Characteristic permanent action due to supported wall self-weight =

$$G_{k,wall} = 2.75\text{m} \times g_{k,wall} = 1.38 \text{kN/m}$$

Supported floor self-weight is taken as

$$g_{k,floor} = 1.00 \text{kN/m}^2$$

Characteristic permanent action due to supported floor self-weight =

$$G_{k,floor} = \frac{1}{2} \times 5\text{m} \times g_{k,floor} = 2.5 \text{kN/m}$$

Total characteristic permanent action =

$$G_k = G_{k,swt} + G_{k,wall} + G_{k,floor} = 3.98 \text{kN/m}$$

Design action due to permanent actions =

$$F_{d,p} = \gamma_{G,ULS} \times G_k = 3.98 \text{kN/m}$$

Characteristic variable action due to supported floor =

$$Q_{k,floor} = \frac{1}{2} \times 5.0\text{m} \times 1.50 \text{kN/m}^2 = 3.75 \text{kN/m}$$

Design action due to variable action =

$$F_{d,q} = \Psi_1 \times Q_{k,floor} = 1.88 \text{kN/m}$$

Modification factors

The designs considered are post-accidental event (i.e., for the residual structure) and the applicable k_{mod} values are, therefore, those applicable to short-term duration loads, as provided by Table NA.1 of the UK NA to BS EN 1995-1-1.

Factor for short-term duration loading and SC1
(BS EN 1995-1-1, Table 3.1) =

$$k_{mod,sht} = 0.90$$

Size factor for depth less than 300mm
(BS EN 1995-1-1, Expression 3.3 and Kerto VTT
Certificate No. 184/03 March 2009) =

$$k_h = \min. \begin{cases} \left(\dfrac{300}{240}\right)^{0.12} = 1.03 \\ 1.2 \end{cases}$$

Lateral stability of the beam factor
(BS EN 1995-1-1, Expression 6.3.3) =

$$k_{crit} = 1.0$$

Bearing factor $k_{c,90}$ for LVL (Kerto VTT
Certificate No. 184/03 March 2009) =

$$k_{c,90 \text{ (LVL)}} = 1.0$$

Bearing factor $k_{c,90}$ for C16
(BS EN 1995-1-1, Expression 6.1.5(3)) =

$$k_{c,90 \text{ (C16)}} = 1.25$$

Load sharing factor considering a two-ply LVL rim beam
(BS EN 1995-1-1, Expression 6.6(2)) =

$$k_{sys} = 1.1$$

Beam bearing detail at support (in accordance with BS EN 1995-1-1, Expression 6.1.5(1))

Figure 12.14: Beam bearing detail

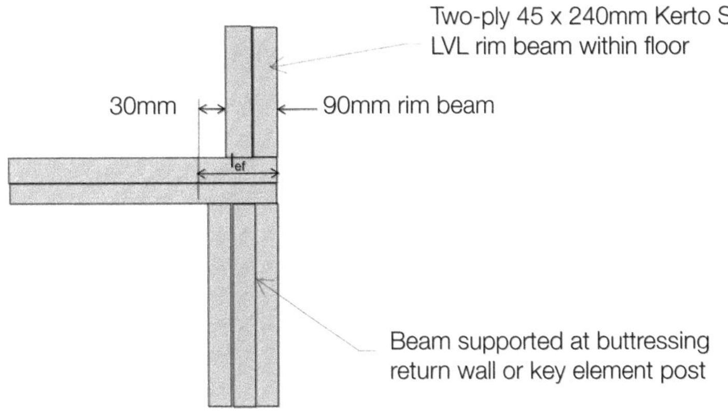

Effective contact area in compression perpendicular to grain = $A_{ef} = l_b \times l_{ef} = 90 \times (b + 30) = 10{,}800\text{mm}^2$

Beam bending strength

Design bending moment for the residual load case = $M_d = \dfrac{(F_{d,p} + F_{d,q})}{8} \times l^2$

$$= (3.98 + 1.88) \times \dfrac{4.59^2}{8} = 15.5\text{kNm}$$

Design bending stress = $\sigma_{m,y,d} = \dfrac{M_d}{W_y} = \dfrac{15.5 \times 10^6}{8.64 \times 10^5} = 17.9\text{N/mm}^2$

Design bending strength = $f_{m,y,d} = \dfrac{k_{mod,sht} \times k_{sys} \times k_h \times f_{m,k}}{\gamma_m}$

$$= \dfrac{0.9 \times 1.1 \times 1.03 \times 44}{1.0} = 44.9\text{N/mm}^2$$

$f_{m,y,d} > \sigma_{m,y,d}$

Bending strength is satisfactory

Beam shear strength

Design bending moment for the residual load case =

$$V_d = (F_{d,p} + F_{d,q}) \times \frac{l}{2}$$

$$= (3.98 + 1.88) \times \frac{4.59}{2} = 13.5\text{kN}$$

Design shear stress
(BS EN 1995-1-1, Expression 6.60) =

$$\tau_{v,d} = \frac{3}{2} \times \frac{V_d}{b \times h}$$

$$= \frac{3}{2} \times \frac{13.5 \times 10^3}{90 \times 240} = 0.94\text{N/mm}^2$$

Design shear strength =

$$f_{v,d} = \frac{k_{mod,sht} \times k_{sys} \times f_{v,k}}{\gamma_m}$$

$$= \frac{0.90 \times 1.1 \times 4.1}{1.0} = 4.06\text{N/mm}^2$$

$$f_{v,d} > \tau_{v,d}$$

Shear strength is satisfactory

Bearing strength

Design bearing force at supports = Design shear force V_d

For the LVL rim beam, design bearing stress =

$$\sigma_{c,90,d} = \frac{V_d}{b \times l_b} = \frac{13.5 \times 10^3}{90 \times 90} = 1.67\text{N/mm}^2$$

Design bearing strength =

$$f_{c,0,d\,(LVL)} = \frac{k_{mod.sht} \times k_{sys} \times k_{c,90\,(LVL)} \times f_{c,90,k\,(LVL)}}{\gamma_m}$$

$$= \frac{0.90 \times 1.1 \times 1.0 \times 6.0}{1.0} = 5.94\text{N/mm}^2$$

$$f_{c,0,d\,(LVL)} > \sigma_{c,90,d}$$

Bearing strength of LVL is satisfactory

For the C16 plates, design bearing stress =

$$\sigma_{c,90,d} = \frac{V_d}{A_{ef}} = \frac{13.5 \times 10^3}{10,800} = 1.25\text{N/mm}^2$$

Design bearing strength =

$$f_{c,0,d\,(C16)} = \frac{k_{mod.sht} \times k_{c.90.C16} \times f_{c,90,k\,(C16)}}{\gamma_m}$$

$$= \frac{0.90 \times 1.25 \times 2.2}{1.0} = 2.48\text{N/mm}^2$$

$$f_{c,0,d\,(C16)} > \sigma_{c,90,d}$$

Bearing strength of plates is satisfactory

Therefore, the two-ply 45 x 240mm LVL rim beam is able to span over the notionally removed external wall following an accidental event.

Case study **14**

Robustness of CLT — notional element removal

This case study indicates the approach for checking the ability of a first-floor CLT wall element to span across an opening created by notional removal of the ground-floor wall and to support the first-floor and roof above.

Geometry of structure (Figure 12.15)

Figure 12.15: Structural details

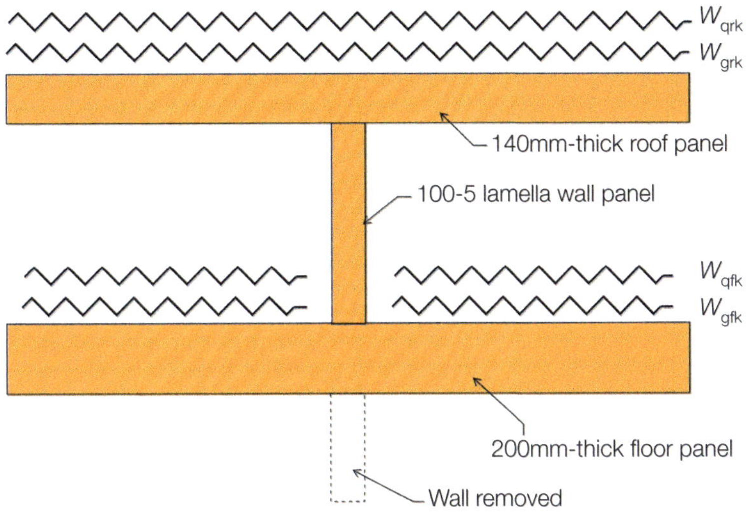

Wall thickness

$$t_w = 100mm$$

CLT roof thickness

$$t_r = 140mm$$

CLT floor thickness

$$t_f = 200mm$$

Floor-floor height

$$F = 2.95m$$

Clear height

$$H = F - t_f = 2.75m$$

Clear span of wall

$$l_c = 2.25 \times h = 6.19m$$

Notional length of bearing

$$b = 200mm$$

Design span

$$l = l_c + 2 \times \left(\frac{b}{2}\right) = 6.39m$$

CLT wall properties
Outer grain spanning parallel to the short length, i.e., Binderholz DQ type panel
Lay-up: 20/20/20/20/20
No vertical panel splits in region under consideration
Properties from Binderholz ETA 06-0009:

$$E_{0,mean} = 12{,}000 \text{N/mm}^2$$
$$G_{mean} = 690 \text{N/mm}^2$$
$$G_{r,mean} = 50 \text{N/mm}^2$$
$$f_{m,k} = 24 \text{N/mm}^2$$
$$f_{R,k} = 1 \text{N/mm}^2$$
$$f_{v,k} = 3.2 \text{N/mm}^2$$
$$f_{c0,k} = 21 \text{N/mm}^2$$
$$f_{c90,k} = 2.5 \text{N/mm}^2$$
$$f_{t0,k} = 14.5 \text{N/mm}^2$$
$$\rho_k = 4.5 \text{kN/m}^3$$

Sum of thickness of inner lamellae

$$t_{inner} = 20\text{mm} + 20\text{mm} = 40\text{mm}$$

$$W_y = \frac{t_{inner} \times H^2}{6} = 50{,}416{,}667 \text{mm}^3$$

Partial safety factors

$$\gamma_{G,ACC} = 1$$
$$\Psi_1 = 0.5$$
$$\gamma_M = 1$$

Actions

$$g_{w,k} = t_w \times H \times \rho_k = 1.24 \text{kN/m}$$
$$g_{f,k} = t_f \times \rho_k + 2\text{kN/m}^2 = 2.9 \text{kN/m}^2$$
$$q_{f,k} = 2.5 \text{kN/m}^2$$
$$g_{r,k} = t_r \times \rho_k + 1\text{kN/m}^2 = 1.63 \text{kN/m}^2$$
$$q_{r,k} = 0.6 \text{kN/m}^2$$

Loaded width

$$w = 5\text{m}$$

Accidental design action

$$\omega = \gamma_{G,ACC} \times g_{w,k} + w \times (\gamma_{G,ACC} \times [g_{f,k} + g_{r,k}] + \Psi_1 \times [q_{f,k} + q_{r,k}]) = 31.64 \text{kN/m}$$

Modification factors

Load duration

$$k_{mod} = 0.9$$

Lateral stability

$$k_{crit} = 1$$

Wall is prevented from buckling by connections to the roof and floor above and below

Bearing factor

$$k_{c90} = 1.9$$

From pro:Holz Volume 1[103]

Bending strength

$$M_d = \frac{\omega \times l^2}{8} = 161.35 \text{kNm}$$

$$\sigma_{m,y,d} = \frac{M_d}{W_y} = 3.2 \text{N/mm}^2$$

$$f_{m,y,d} = \frac{k_{mod} \times f_{m,k}}{\gamma_M} = 21.6 \text{N/mm}^2$$

$f_{m,y,d} > \sigma_{m,y,d}$ therefore satisfactory

Shear strength

$$V_d = \omega \times \left(\frac{l}{2}\right) = 101.04 \text{kN}$$

$$\tau_d = 1.5 \times \left(\frac{V_d}{t_{inner} \times H}\right) = 1.38 \text{N/mm}^2$$

Making no allowance for k_{cr} as using inner lamellae only

$$f_{v,d} = \frac{k_{mod} \times f_{v,k}}{\gamma_M} = 2.88 \text{N/mm}^2$$

$f_{v,d} > \tau_d$ therefore satisfactory

Bearing check

Bridging allowance

$$b_r = 30\text{mm}$$

Bearing width

$$w_b = t_w + 2 \times b_r = 160\text{mm}$$

$$\sigma_{c,90,d} = \frac{V_{e,d}}{w_b \times b} = 3.16\text{N/mm}^2$$

$$f_{c,90,d} = \frac{k_{c90} \times f_{c90,k}}{\gamma_M} = 4.75\text{N/mm}^2$$

$f_{c,90,d} > \sigma_{c,90,d}$ therefore satisfactory

A 100mm-thick five lamella wall, grain parallel to short edge, with no panel splits, can span over the notional removal length. Capacity of wall-to-floor brackets to transfer floor load in tension to wall, will also need to be checked.

Case study 15

Robustness of CLT — tying

This case study indicates the approach for generating discrete internal horizontal and vertical ties in CLT construction, where proprietary angle brackets connect the walls to the substructure and raised-thread self-tapping screws connect the CLT floor slabs to the supporting walls. The calculation demonstrates the required centres for these brackets and screws to develop the required tie forces.

Geometry of structure (Figure 12.16)

Figure 12.16: Structural details

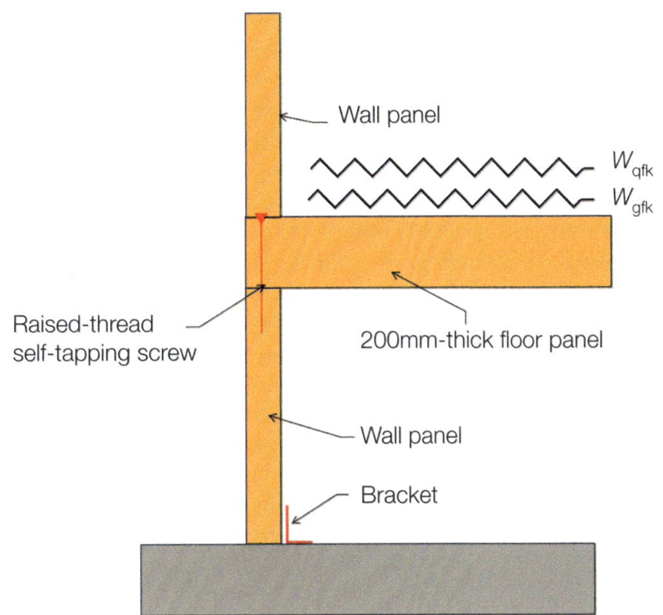

CLT floor thickness

$$t_f = 200mm$$

Floor span

$$L = 5m$$

CLT properties

$$\rho_k = 4.5kN/m^3 \ (450kg/m^3)$$

For calculation of self-weight
Note that for the calculation of screw capacity it is typically taken as $350kg/m^3$

Partial safety factors

$$\gamma_{G,ACC} = 1$$
$$\Psi_1 = 0.5$$
$$\gamma_M = 1$$

Actions

$$g_k = t_f \times \rho_k + 2\text{kN/m}^2 = 2.9\text{kN/m}^2$$

$$q_k = 2.5\text{kN/m}^2$$

Vertical action on wall

$$G_k = \frac{L \times g_k}{2} = 7.25\text{kNm}$$

$$Q_k = \frac{L \times q_k}{2} = 6.25\text{kNm}$$

Distributed internal tie

$$T_i = 0.8 \times (g_k + \Psi_1 \times q_k) \times L = 16.6\text{kN/m}$$

Linear vertical tie

$$T_v = (G_k + \Psi_1 \times Q_k) = 10.38\text{kN/m}$$

Modification factors

Load duration

$$k_{mod} = 0.9$$

For residual structure after an accidental event refer to UK National Annex to BS EN 1995-1-1, Table NA.1

Screw capacity

From separate calculation, shear capacity of ø8 x 260 Heco Topix Flangehead screws into the narrow edge of a CLT panel

$$F_{v,Rk} = 2.3\text{kN}$$

$$F_{v,Rd} = k_{mod} \times \left(\frac{F_{v,Rk}}{\gamma_M}\right) = 2.07\text{kN}$$

As $n_{ef} = n$ for screws in shear in CLT

$$\text{number of screws} = \frac{T_i}{F_{v,Rd}} = 8 \text{ per metre length}$$

So provide 10 screws per metre, i.e., screws at 100mm centres

Bracket capacity to foundation

From catalogue, tension capacity of Rothoblaas WKR13535 brackets, fully nailed

$$R_{1k,timber} = 28.3\text{kN}$$

$$R_{1d,timber} = k_{mod} \times \left(\frac{R_{1k,timber}}{\gamma_M} \right) = 25.47\text{kN}$$

Using Rothoblaas VIN-FIX M12 x 195 anchor

$$R_{1k,concrete} = 20.5\text{kN}$$

$$R_{1d,concrete} = \frac{R_{1k,concrete}}{\gamma_M} = 20.5\text{kN}$$

Take concrete value as smaller

$$\text{bracket centres} = \frac{R_{1d,concrete}}{T_v} = 1.98\text{m}$$

Provide brackets to foundations at min. 2m centres

A calculation for axial capacity of the floor–wall screws would also be required

13 Design of structural alterations to existing buildings

13.1 Introduction to structural alterations to existing buildings

Engineers in the UK have the opportunity to work with existing buildings, many of which are historic or have architectural value. Owing to the UK's planning rules and appreciation of older buildings, there has been a tradition of restoration and alteration alongside demolition and rebuild.

Much of an engineer's workload is at the domestic level, extending or opening up houses and creating attic conversions. The traditionally-built masonry houses in which many people live have proved to be highly adaptable as needs have changed over the years.

Bigger buildings have often had different uses over their lifespan, with industrial buildings converted to housing, single-family units turned into flats, and residential buildings transformed into restaurants.

Throughout the 20th century, materials other than masonry and timber were introduced in the construction of larger buildings, with steel and concrete becoming more typical. In the later part of the 20th century, prefabrication of structural elements became more popular to enable fast construction. As materials became stronger, the amount of structure required in a building decreased.

There are many challenges facing the engineer; an understanding is necessary of the existing building, its structure and materials, its previous uses and how it has been altered over the years.

Alterations to existing buildings are a high-risk activity and it is vital that the engineer constantly considers this.

13.2 Statute and legislation

The first step is to understand current legislation on alterations to existing buildings.

The government has produced the following important documents regarding structural engineering:

- *The Building Regulations 2010*[26] (Section 13.2.1)
- *The Building Regulations 2010. Approved Document A: Structure*[25] (Section 13.2.2)
- *Review of international research on structural robustness and disproportionate collapse*[45] (Section 13.2.3)
- *Party Wall etc. Act 1996*[104] (Section 13.2.4)
- *The Construction (Design and Management) Regulations 2015*[13] (Section 13.2.5)
- *Building Safety Act 2022*[2] (Section 13.2.6)

13.2.1 The Building Regulations 2010
In England and Wales, The Building Act 1984[47] is the primary, enabling legislation, under which secondary legislation such as the Building Regulations is introduced.

It empowers the Secretary of State to make regulations for the purpose of:

- Securing the health and safety of persons in and about buildings
- Furthering conservation of fuel and power
- Preventing waste and undue consumption, misuse or contamination of water

The Building Regulations made under this Act prescribe notification procedures that must be followed when starting, carrying out and completing building work, and set out minimum standards of design and construction.

Regarding the Statute (the law), there are three main areas to consider:

- Clause 5. Material change of use — where the use of a building is changed, such as converting apartments into a hotel. There are specific definitions for material change of use and the engineer should consult the Statute
- Clauses 3 and 4. Building work — how to implement changes to an existing building and when it must comply with Schedule 1. Two of the definitions of 'building work' are 'material change of use' and 'material alteration of a building'. Another definition is the 'extension of a building'
- Schedule 1 (Structure). This is the minimum requirements for a building, including a note stating that the building must be resistant to disproportionate collapse, A3. The most important parts of this document are:

  "Clause 4.(3). Building work shall be carried out so that after it has been completed —

  Clause 4.(3).a — Any building which is extended or to which a material alteration is made complies with the applicable requirements of Schedule 1 or, where it did not comply with any such requirement, is no more unsatisfactory in relation to that requirement than before the work was carried out."

The part stating *"is no more unsatisfactory in relation to that requirement than before the work was carried out"* is important. The Statute does not allow an engineer to make an existing building any worse — it is the law and the basis of this chapter, and may actually aid an engineer with their design.

However, for some building works and some categories of material change of use, the entire structure of the altered building is expected to comply with Schedule 1 of the Building Regulations once the works have been undertaken.

In Scotland, the requirements are more onerous. *The Technical Handbook*[105] states that:

"...the building as converted shall meet the requirements of this standard in so far as is reasonably practical, and in no case be worse than before the conversion."

Therefore, in Scotland, the engineer is looking to make the existing building compliant with the current regulations, 'as far as is reasonably practical' in addition to the requirement in England and Wales.

> It is recommended that the Building Regulations are consulted, with reference to understanding the definitions of 'building work' and 'material change of use' set out in Clauses 3, 4, 5 and 6 (Figure 13.1).

When making alterations to existing buildings, it is important to consider that many will not comply with the Building Regulations even before any alterations are made, e.g., the Ronan Point collapse in 1968. This was a critical event that changed the way UK engineers considered robustness and put the UK ahead of other countries. In 1970 the Building Regulations were amended and in 1972 CP 110, the unified code of practice for concrete[106] was revised. It was in these publications that the term 'robustness' was formally used for the first time.

Therefore, any building constructed before 1970 will not have been designed for robustness, although an earlier building may be robust due to the way it is structured. The significance of Ronan Point is that engineers must have full understanding of the building structure when designing alterations to pre-1970 buildings.

Engineers should also be wary of blindly following prescriptive rules within the approved documents, and look to understand the actual structural performance of the building before and after any alterations.

It should also be noted that undertaking alterations to a building which does not make the building 'more unsatisfactory' is not good enough for buildings that were considered to be dangerous before alterations were undertaken. A seemingly logical argument, using the relevant clauses in the Building Regulations, is no excuse for taking a dangerous building, doing structural alterations, and leaving it as dangerous as it was previously.

 *BCA Technical Guidance Note 21: The building regulations 2010 — England and Wales requirement A3 — disproportionate collapse*[107]

 *BRE Digest 366 Parts 1–4 — Structural appraisal of existing buildings, including for a material change of use*[108]

Figure 13.1: The Building Regulations 2010 — material change of use flowchart

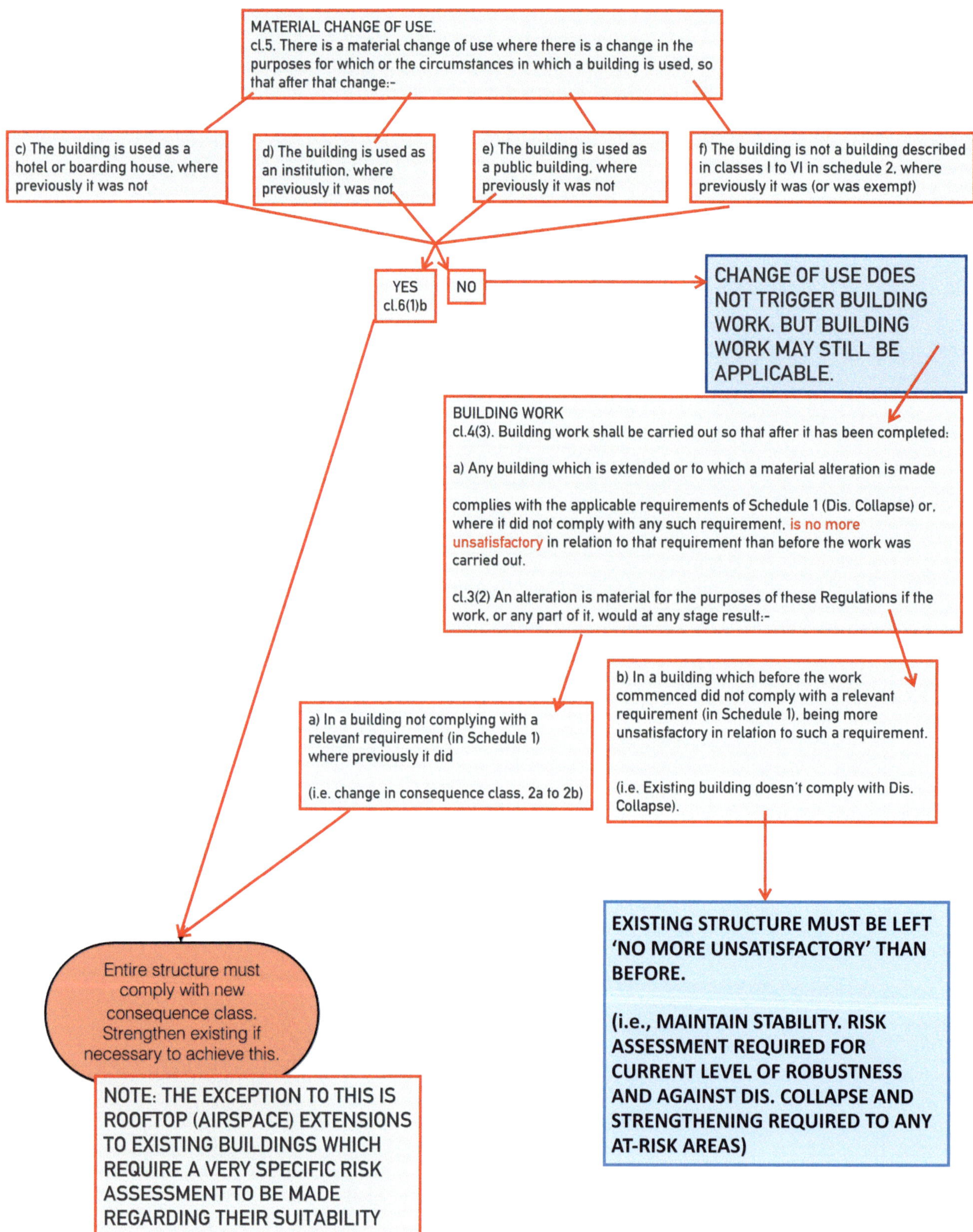

13.2.2 The Building Regulations 2010. Approved Document A: Structure

The Building Act also sets out the legal status of the approved documents, which provide general guidance on how building, design and construction can comply with building regulations.

The approved document relating to structure, is Approved Document A, but it is important to revert to the actual Building Regulations because Approved Document A does not cover alterations; it only covers new-build structures. However, Clause A3 sets out robustness and disproportionate collapse guidance, all of which are applicable when an existing building is altered or extended. A3 is Statute and is part of Schedule 1 of the Building Regulations, along with A1 and A2.

The Statute DOES cover alterations and extensions to existing buildings.
Approved Document A DOES NOT cover alterations and extensions to existing buildings.

13.2.3 *Review of international research on structural robustness and disproportionate collapse*

This is a comprehensive document and should be read in its entirety because it provides an in-depth review of robustness, and also discusses existing buildings. It is also discussed in Section 4.10 of this *Guidance*.

The *Review of international research on structural robustness and disproportionate collapse* DOES cover alterations and extensions to existing buildings.

13.2.4 Party Wall etc. Act 1996

"An Act to make provision in respect of party walls, and excavation and construction in proximity to certain buildings or structures."

The Party Wall Act ensures that robustness and stability of adjacent properties are considered within the design. Consideration should be given at conceptual design stage 2. Early understanding of party wall issues will often dictate the proposed solution, and details should be submitted to the party wall surveyor at the earliest opportunity for approval.

Important points to consider are the rules on:

* Excavating within 3m of an adjacent structure (Figure 13.2)
* Excavating below an adjacent structure within 6m (Fig. 13.2)
* Making alterations to an existing party or boundary wall
* Extending or reducing in height an existing party or boundary wall
* Constructing special (reinforced) foundations under a party wall

Figure 13.2: Party Wall Act 3m and 6m notices[109]

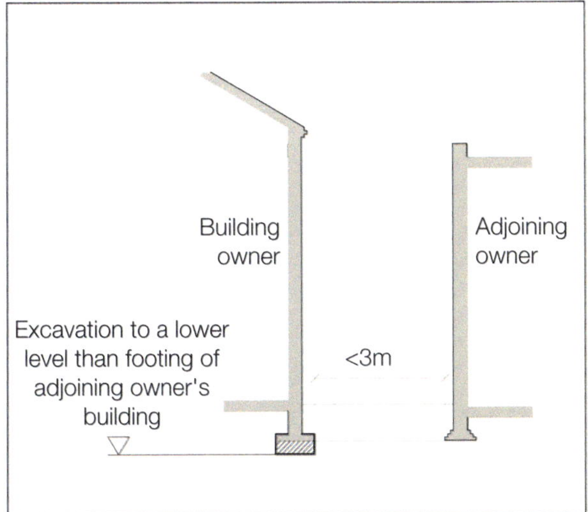

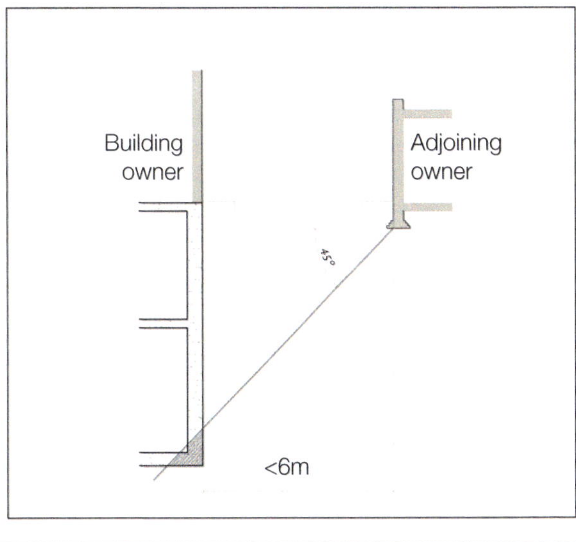

While there is a lot of guidance on party wall compliance, it is recommended that this Act is read in its entirety.

"Failure to agree party wall awards can lead to delays to the start of work on site."[110]

In addition to the Party Wall etc. Act 1996, the nature of building ownership will impact on whether works to a building are permitted. With respect to apartment blocks or multiple-occupancy structures, the freeholder maintains the right to alter and extend the premises and this also covers the ability to commission any investigation and testing work to the existing structure (with the exception of leaseholder apartments).

Where a leaseholder seeks to make alterations within their property, permission of the freeholder will typically be required.

13.2.5 The Construction (Design and Management) Regulations 2015

The structural engineer must fully understand the designer role under CDM Regulations. Clause 9 sets out the duties of designers:

"Clause 9.4. A designer must take all reasonable steps to provide, with the design, sufficient information about the design, construction or maintenance of the structure, to adequately assist the client, other designers and contractors to comply with their duties under these Regulations."

The requirement to consider how a structure will be constructed is more important with existing structures and their alterations. The structural engineer should not only have knowledge of the permanent works design, but also an appreciation of the envisaged construction method, sequence and temporary works requirements, where applicable, with each element being communicated under the requirements of the CDM Regulations. BS 5975[111] (temporary works) also reiterates this point.

"Clause 8.3.1. Permanent works designers should address the buildability of the permanent works and identify and make provision for any temporary works and temporary conditions required by their design and their assumed method of construction."

13.2.6 Building Safety Act 2022

The Building Safety Act covers some existing buildings, including when works are undertaken to existing higher-risk buildings (HRBs) or when a higher-risk building was constructed and occupied before the Act was published.

The design engineer needs to be aware of some key clauses in the Act:

"Clause 71(2). A higher-risk building is 'occupied' if there are residents of more than one residential unit in the building."

"Clause 76(1). This section applies if any of the following works are carried out:

Clause 76(1)a. The construction of a higher-risk building (new-build structures)
Clause 76(1)b. The creation of additional residential units in such a building (lateral and vertical extensions or change of use)
Clause 76(1)c. Works to a building that cause it to become a higher-risk building (vertical extensions which tip a building into seven storeys or 18m or greater)"

For occupied HRBs there will also be a duty to apply for a building assessment certificate.

"Clause 79(1). This section applies where the regulator directs the principal accountable person for an occupied higher-risk building to apply to the regulator for a building assessment certificate in relation to the building."

The assessment of building safety risks falls under Clause 83.

"Clause 83(1). An accountable person for an occupied higher-risk building must as soon as reasonably practicable after the relevant time assess the building safety risks as regards the part of the building for which they are responsible."

Therefore, for existing HRBs an assessment of the existing structure, its structural performance and any risks will need to be made.

"Clause 83(2). Further such assessments must be made:

Clause 83(2)a. At regular intervals."

Which, it is assumed to mean, that regular assessments will be required for higher-risk buildings on a regular basis to mitigate any structural risks that may occur.

"Clause 84(1). An accountable person for an occupied higher-risk building must take all reasonable steps for the following purposes:

Clause 84(1)a. Preventing a building safety risk materialising as regards the part of the building for which they are responsible.
Clause 84(1)b. Reducing the severity of any incident resulting from such a risk materialising."

The responsibility of the building integrity is placed firmly back with the accountable person, who must maintain and document clear evidence that all building risks are mitigated, and where they do occur, are resolved. The accountable person is likely to rely on evidence provided by the structural engineer, in the form of regular assessments on the building, and while there will always be commercial pressures placed upon the structural engineer in regard to any remedial works flagged during an assessment, it is important to remember points 1 and 2 of the IStructE's code of conduct[112]:

1. Act with integrity and fairness and in accordance with the principles of ethical behaviour
2. Have regard to the public interest as well as the interests of all those affected by their professional activities

13.2.6.1 Risk assessment of existing HRBs

Assessment of existing HRBs is in the form of a risk assessment for structural failure and fire spread, and should be included in the safety case report for each building. The risks to be considered are not just specific to disproportionate collapse, but must include all risks which could affect structural performance.

While it might sound obvious, a risk assessment is 'risk-based' and not 'prescriptive'. A prescriptive approach would be one that is code-compliant. This is easily carried out for a new building, but much more challenging in existing buildings, especially those that are occupied.

A risk-based approach becomes a more powerful tool when dealing with existing building stock. Each building is different in size, shape, height, occupancy and materials and therefore there are different risks for each. There will not be an overall prescriptive approach; each building should be assessed on its own merits, with the structural engineer carrying out appropriate investigation and analysis, to understand what the risks to that building are, and to determine how these risks can be mitigated and removed.

A holistic view of the structure is required and, where mitigation measures are not adopted for specific risks, clarification needs to be demonstrated on how this decision was reached. A safety measure omitted based on cost alone is not permitted.

Structural engineers will quickly need to become comfortable with terms such as:

- As low as reasonably practicable (ALARP)
- Risk matrix
- Cost benefit analysis (CBA)
- Gross disproportion

At present, the IStructE's *Manual for the systematic risk assessment of high-risk structures against disproportionate collapse*[19] gives guidance on this.

For each building, engineers will be expected to demonstrate clearly the risks, proposed mitigation measures and associated risks from implementing those mitigation measures as part of any CBA.

This is echoed in Section 6.4 of the '*Manual for the systematic risk assessment of high-risk structures against disproportionate collapse*'.

The Health and Safety Executive (HSE) guidance to support a risk assessment process includes:

- Principles and guidelines to assist HSE in its judgements that duty-holders have reduced risk as low as reasonably practicable[113]
- Assessing compliance with the law in individual cases and the use of good practice[114]
- Policy and guidance on reducing risks as low as reasonably practicable in design[115]
- Principles for the cost benefit analysis in support of ALARP decisions[116]
- Reducing risks, protecting people: HSE's decision-making process[117]

13.2.6.2 Works to existing buildings

When working on existing buildings, the structural engineer must check if the building might be an HRB, or if the works designed put the building into the HRB category. The Building Safety Act was created in response to a tragedy, and people expect to feel safe in their homes. Structural engineers are relied upon to ensure that safety.

13.3 Understanding the existing building

Before an engineer produces a fee proposal and scope of works for a project, it is wise to look at the existing structure and ask some important questions:

- How is the existing structure acting?
- What is needed to make the proposed scheme work?
- What was the original design life of the building?
- What is the actual condition of the building?
- How much residual capacity is there in the building to enable safe alterations?

Table 6.1 of IStructE's *Appraisal of existing structures*[118] provides a timeline to understand the main forms of construction throughout history. A simplified version can be seen in Figure 13.3. Understanding where the building lies on this timeline will give the engineer a basic appreciation of expected structural form, construction and typical design codes used.

Figure 13.3: Materials and loading timeline

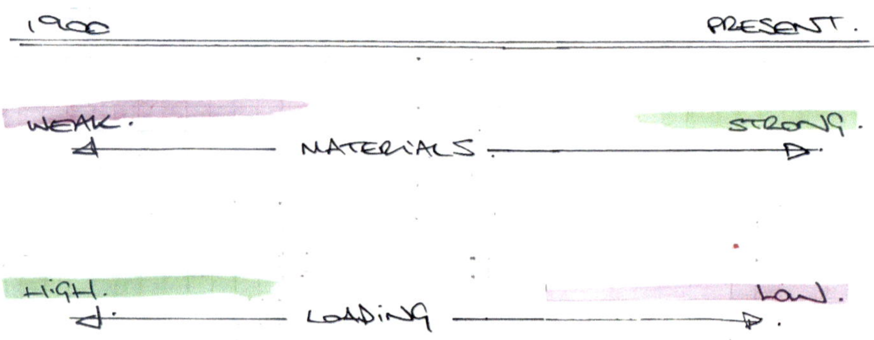

The age and form of construction reveal much about a building. The older the building the weaker the materials (lime mortar within brickwork, rebar, steel, etc.). Some buildings may not even have proper foundations. A review of the old codes will reveal the recommended material strengths and loading adopted at the time of construction. Before the fee proposal is prepared, the engineer should fully understand the structure and how the current vertical and lateral load paths are resolved.

When assessing buildings that have been built more recently, the engineer may have access to design drawings and calculations. However, these always need to be treated with caution; buildings may have deteriorated such that their structural performance is impacted and the building may be at the end of its design life, even if it appears relatively modern.

Engineers should also become familiar with the concept of 'Reliability Index'. ISO 13822[119,] the international standard for the assessment of existing structures, describes the target reliability level as the required performance level for the structure, the reference period and the possible failure consequences. BS EN 1990 Annexes B and C discuss the management of reliability and the theory behind it.

Once there is a clear picture of the existing structure it is important to look at the proposed scheme, and specifically how that affects the existing structure:

- Are walls being removed that are thought to be loadbearing or stability walls?
- Where do those existing load paths go?
- Are new foundations required?
- How do the proposed works affect any party walls?
- How is the contractor going to build it and is access to and within the site limited in any way?

Frequently these questions will guide the engineer to a scheme, or at the very least highlight a key risk on the proposed scheme, which the client and/or architect may not have considered. Sketching this is often a good way to communicate those risks to the client at the fee proposal stage.

 *ISO 13822 — Bases for design of structures — Assessment of existing structures*[119]

13.3.1 Visual condition survey

A visual investigation and assessment of the construction and condition of a building would not usually include advice on value or costs. The exposure and testing of services are also not typically covered. Any advice on the cost of repairs should be sought from a building/quantity surveyor where applicable. It will not usually include the cost of exploratory or intrusive work.

The visual condition survey report will probably only include reference to visible defects and guidance, as appropriate, on maintenance and remedial measures. The report may recommend that elemental or specialist investigations are undertaken or other specialist advice obtained relating to specific issues[120].

The visual condition survey is the first impression and assessment by the engineer of the structure and its strengths and weaknesses. It shows how it is behaving, and will suggest how it is likely to behave under proposed works. This is key, not only from an engineering and information gathering exercise, but many insurance companies and financial lenders will want satisfaction that the existing building is in good condition, and may require a visual condition survey to confirm.

While the client and architect are envisaging the proposed works, it is vital that the engineer gathers and records as much information as possible about the existing structure:

- How old is the structure?
- What was the period/year of construction and to what codes (where applicable) was it designed?
- What is the primary vertical framing system? Loadbearing walls or framed structure?
- Is the building of cavity wall or solid wall construction? Infill or primary element?
- How is overall stability currently achieved? Which elements are resisting vertical and lateral forces?
- What form of foundations are likely, due to its age and form of construction?
- Which areas of the structure can be accessed for any future investigation and testing works — only communal areas of apartment blocks?
- Is the external facade surrounded by hardstanding (pavements, parking, etc.) or are there suitable areas for trial pits to be created?
- Are there any areas of restricted access which may need to be inspected at a later date?
- Does the structure contain any transfer structures within it?
- What is the general condition of the structure in relation to its age? Has it settled, moved, undergone any remedial works? If so, why?

All buildings move and the engineer should be looking for any cracks, dips, slips and movement that may be affecting the structural performance of the property. It is vital that the engineer looks at these first, rather than the proposed end-product, as these issues can dramatically affect the capacity of the structure and ultimately the proposed scheme.

13.3.2 Investigation and testing

Where testing and investigation are required, the IStructE's *Appraisal of existing structures* is a useful publication. It is a comprehensive document, and details both general and more specific testing regimes.

It is recommended to take as many samples as possible; an ideal range would be a minimum of 6–10 samples. Testing is a very low-cost activity in relation to the overall project cost, but yields very important information, which is often needed not only by the design engineer but also any future insurers of the structure, who will require assurance that the building has capacity for any change in loading or use.

The timing of when to undertake investigation and testing should also be considered. Typically it should be proposed within stage 2 design (pre-planning) and complement any feasibility report which has identified areas of overstress and testing requirements.

The risk to the client by commissioning testing at this point is that they fail to achieve planning consent. The decision on planning (once the application has been submitted) may take several months, which presents the opportunity for investigation and testing to take place without affecting programme.

Clients should be made aware of the effect on programme should they wish to push investigation and testing to the post-planning stage.

Typical permissible stresses adopted are:

Masonry
- $0.42N/mm^2$ for compressive strength (x1.5) as a loadbearing enhancement factor under point loads[121]. The enhancement factor is commonly adopted throughout all codes (including Eurocodes)

Steel (in tension). Values differ for steel in compression and shear and a check of previous codes should be made:
- $124N/mm^2$ (1879–1912)
- $127N/mm^2$ (1912–1961)
- $151N/mm^2$ (1962–1969)

Timber
- See Section 6.3.2 of *Appraisal of existing structures* and Table 13.1 of this *Guidance*

Table 13.1: Species of timber used over time

Approximate period	Origin	Species
1780–1830	UK	Oak and other local hardwoods Some softwoods (not very common)
	Scandinavia/Russia	Redwoods/whitewoods
	Canada/US	Douglas fir, pitch pine
1880–1930	Baltic states	Redwoods/whitewoods
	Canada/US	Spruce, Douglas fir, pitch pine
	UK	Scots pine, oak (in selected applications)
1930 onwards	Scandinavia/Russia	Redwoods/whitewoods
	Tropical regions	Tropical hardwoods
	Canada/US	Douglas fir, pitch pine, yellow pine, western hemlock Spruce/pine/fir (from 1970)

- BM TRADA Wood Information Sheet (WIS) 1-34[122] states:
  "Moisture ingress allows timber to become wet and decay. This is not uncommon, particularly in older structures. Since timber is an organic material, prolonged exposure to high levels of moisture can lead to decay by fungi. Timber will not decay if the moisture content is less than 20%. Hence good design practice is to achieve and maintain conditions where wood is kept dry (that is the moisture content is less than 20%)."

  and:

  "You may need to evaluate the loadbearing capacity of structural timbers. This requires in situ visual strength grading. When assessing the strength of softwood, a qualified grader will follow the specification in BS 4978 Visual strength grading of softwood. For hardwoods, BS 5756 Visual grading of temperate hardwood or BS EN 16737 Visual strength grading of tropical hardwood will be followed. Record the timber species and grading features such as knots, slope of grain and other characteristics in order to assign a visual strength grade.

  "Then, from the species/grade combination, assign the strength class using BS EN 1912 Structural timber. Strength classes. Assignment of visual grades and species.

  "While performing in situ grading, record moisture contents; a detailed moisture content survey may be required.

  "WIS 4–7 'Timber strength grading and strength classes' describes the system for grading timber."

Concrete
- See Section 6.5 of *Appraisal of existing structures*. *"Little information is obtainable about concrete made earlier than 1909, when the London County Council (LCC) figure for working load design stress was…"*
- $2.5N/mm^2$ (1909)
- $4.5N/mm^2$ (1921)
- $5.2N/mm^2$ (1934)
- Historic concrete: background to appraisal[123] also provides a compressive table of concrete strengths in relation to codes of construction across the 20th century

Rebar
- $110N/mm^2$ (1911)
- $124N/mm^2$ (1934–1965)
- $138N/mm^2$ (1969)

It is worthwhile conducting a desktop study or quick assessment on the material to be tested to gauge the values expected from testing, rather than blindly accepting the results. The test results should be used to confirm design and assumptions rather than dictating the limitations of design. By anticipating the results, the engineer will be better placed to engage a retest or specify further testing where unexpected results are encountered.

Once the results are obtained the data should be analysed and adjusted for use in design. Clause 5.8 of CIRIA Report 111 covers sampling and testing of iron and steel, but the principles can also be applied across a range of materials:

"The average strength of the samples should be considerably higher than the design ultimate strength (or 95% confidence limit), because the scatter of results would be considerable. The samples should also be taken from representative parts of the structure.

"For 6 to 10 samples, the 95% confidence level is in theory the average strength minus two standard deviations (SD) of the test results…around six samples from a series of similar elements usually suffice, provided that the material chosen is reasonably representative of the quality of the remaining structure, including its likely defects."

Case study 16 details an example of masonry testing.

Case study 16

Masonry testing example

In this study an average mean compressive strength of 13.9N/mm^2 was tested — this is not the value to be used. Adjusting for variance and applying two standard deviations results in an average characteristic compressive stress for the masonry of 3.3N/mm^2.

A factor of safety needs to be applied to this value (typically between 2.0 and 3.0 dependent upon the condition of the masonry), which results in the design permissible stress to be used (1.65 to 1.10N/mm^2 in this instance). Note, characteristic loads are to be used where a factor of safety has been applied as a material factor (as per previous code design practice) (Figure 13.4).

It is also worth noting the reliability index value of the material here, too, according to Clause 2.2, Annexes B and C of BS EN 1990. The reliability index is a measure of how much safety lies within a structural element.

Appendix 10 of IStructE's *Appraisal of existing structures* provides evidence of using reduced partial safety factors when designing alterations to existing buildings. However, it mentions nothing of recalibrating the reliability index value and should therefore be used with caution, as the reduction in partial safety factors impacts the inherent safety of the structure. BRE Digest 366[108] discusses this issue further.

Figure 13.4: Masonry testing example

Sample ref., N	Compressive strength, N/mm^2	Variance, σ
1	8.8	-5
2	18.7	5
3	15.6	2
4	7.9	-6
5	8.8	-5
6	13.3	-1
7	22.8	9
8	15.0	1
9		
10		

$$\text{Mean} = \frac{sum}{N} = \boxed{13.9} \quad \text{N/mm}^2$$

$$\text{Variance} = \frac{N_1{}^2 + N_2{}^2 + N_3{}^2 \dots}{N-1} = \boxed{27.8} \quad \text{N/mm}^2$$

$$\text{Standard deviation, SD} = \sqrt{Variance} = \boxed{5.3} \quad \text{N/mm}^2$$

1st SD stress. 68% of values are within 1x SD of the mean $= \boxed{8.6}$ N/mm^2

2nd SD stress. 95% of values are within 2x SD of the mean $= \boxed{3.3}$ N/mm^2

Reliability index $= \boxed{2.6}$

13.3.3 Specific risks when reviewing an existing building

When working on existing buildings it is important to understand how buildings were designed and constructed in the past.

13.3.3.1 Assumptions about connections in steel frames

Historic designs of steel frames to BS 449, first published in 1932[124], adopted one of three design approaches:

- Simple design
- Semi-rigid design
- Fully-rigid design

13.3.3.2 Common risks

There are some common risks that need to be considered across all codes for different materials:

- The use of high-alumina cement between 1950 and 1970
- Punching shear in 1970s codes
- Use of mild steel reinforcement
- Change from serviceability design to ultimate limit state design (timing of change was different for all materials)
- Switch between dead and live loads across codes when taking partitions into account

13.3.3.3 Limit of investigation

Material sampling and investigation only provides a very narrow view of the structure. Any assessment should be made in conjunction with a visual condition survey — examining the condition of existing masonry stock throughout the building, to determine if the sample locations specified represent a fair general condition of the masonry, together with a historic review of the building.

A review of historic maps will allow an overlay of the structure to be made across the ages, and can identify changes in the building shape and sometimes spotting where extensions have been constructed or demolished. On large schemes it might be the case that the entire structure is formed of four or five older structures connected together. In this instance, sampling of material testing may target material from each era of construction to verify overall suitability of the structure.

Where multiple forms of structure have been identified it would also be prudent to undertake an unexploded ordnance survey (UXO). In London, for instance, the building may have been part of a WW2 bombing blitz, which may have prompted the partial rebuild and change between building forms.

13.3.3.4 Building methods across the ages and bespoke housing

The Victorian era saw a boom in masonry construction with the steel industry progressing the use of cast and wrought iron in the latter part of the 1800s. As masonry construction reached its practical limits, construction evolved to see the introduction of steel and composite floors, to create longer spans and bigger buildings. Foundations were often still quite shallow, steel connections were often saddle plates or nominally pinned, and overall stability still relied on the self-weight of large masonry walls.

From 1900 onwards the first building codes started to appear. These rapidly evolved through the 1930s, as the steel industry transitioned from cast iron to wrought iron and the inclusion of high-tensile steel.

After the 1940s, following WW2, as steel was in scarce supply the construction industry turned to reinforced concrete as an alternate means of construction, and its applications were adopted throughout the industry. Robustness and disproportionate collapse were not yet a main design consideration, with panelised systems becoming very popular as the industry looked to construct taller and leaner buildings.

In addition, non-traditional and pre-fabricated housing started to emerge, such as:

- Boot pier and panel cavity houses
- Cornish unit houses
- Woolaway housing
- Stent houses
- Underdown and Winget housing
- Dorlonco steel-framed housing
- No-fines concrete housing
- Reema large-panel dwellings

Each of these has its own unique form of construction and lead up to the Ronan Point collapse in 1968. When working with existing buildings, it is recommended that the structural engineer develops an appreciation for the country's historic construction processes, codes and different methods of construction, as this will aid understanding of the existing structure to a much greater degree.

13.4 Internal alterations and refurbishments

Internal alterations to buildings may have an affect on the robustness of the structure under consideration.

13.4.1 Removing loadbearing walls
The bulk of internal alterations tends to be local isolated elements, with clients wishing to remove existing walls to open up space or reconfigure a floor entirely (Figure 13.5). The solution for each arrangement will generally fall into one of three approaches:

- A beam positioned on padstones
- A portal frame
- A box frame

Figure 13.5: Existing wall and foundation loads

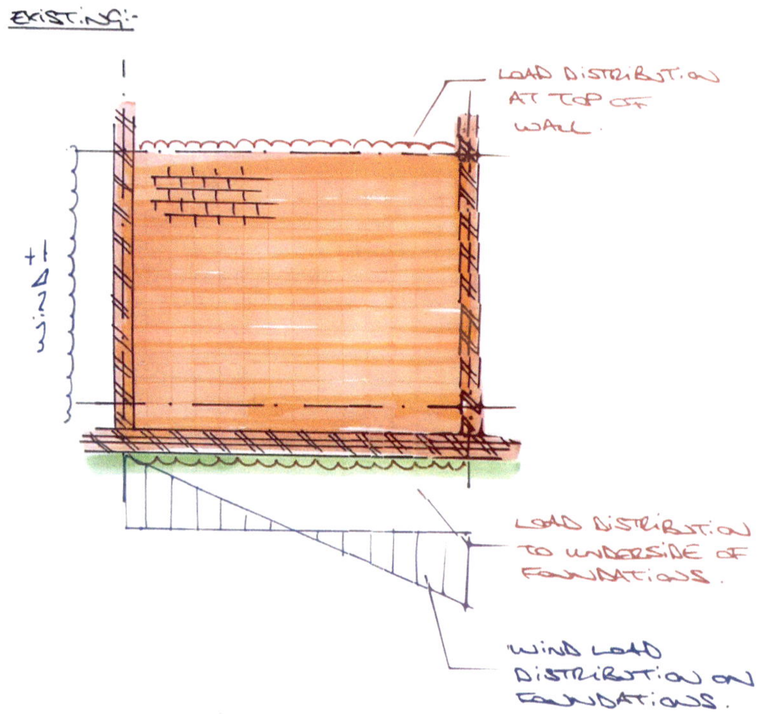

The choice will depend on a number of factors. Typical examples include:

- A small opening within a wall (e.g., a door) (Figure 13.6)
    - Lintel design and immediate support requirements of lintel at bearing
    - Check the remaining wall and foundations (or structure under) for resistance against now concentrated vertical loads, and reduction in overall wall resistance where the wall is providing lateral resistance
    - Other openings within the wall should be considered, which now create small concentrated piers restraining point loads generated by the lintels. These piers can be easily overstressed

Figure 13.6: Opening with lintel and change in foundation loads

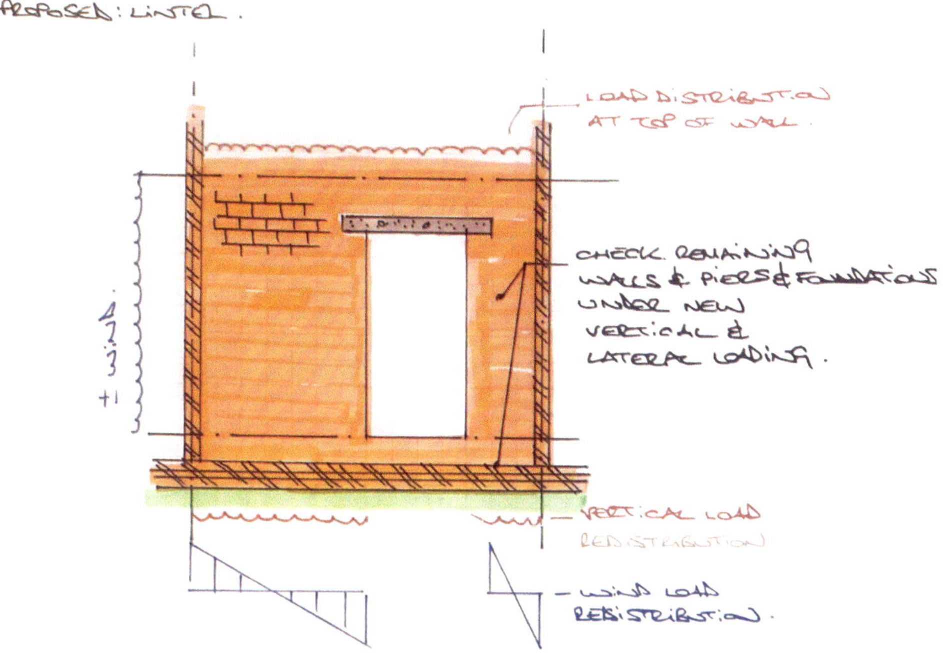

- Large portions or whole walls removed
    - This has the potential to change both the vertical and lateral load paths

Adding a portal frame (Figure 13.7) may solve the vertical and lateral stability issues, but this requires new pad foundations. Check if these can be constructed centred under the columns or if a box frame is required (Figure 13.8). The latter would act to maintain the vertical load paths along the base of the old wall.

Figure 13.7: Proposed portal frame solution

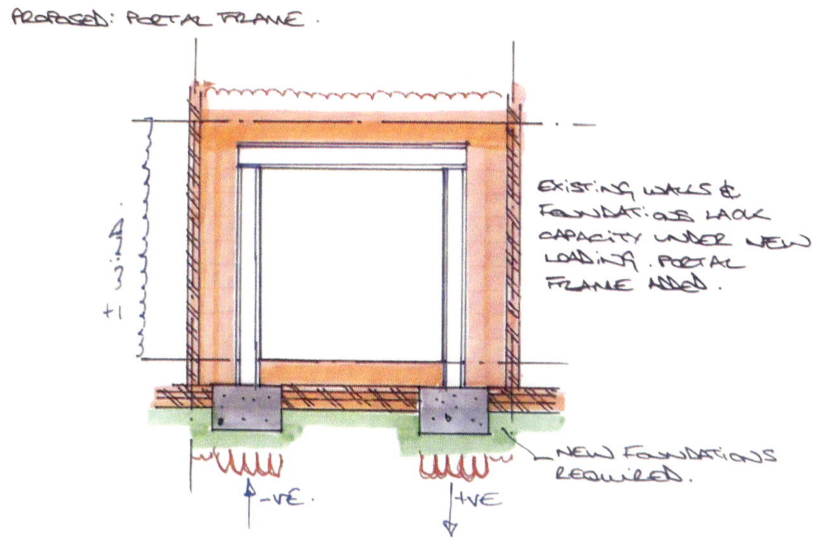

Figure 13.8: A box frame

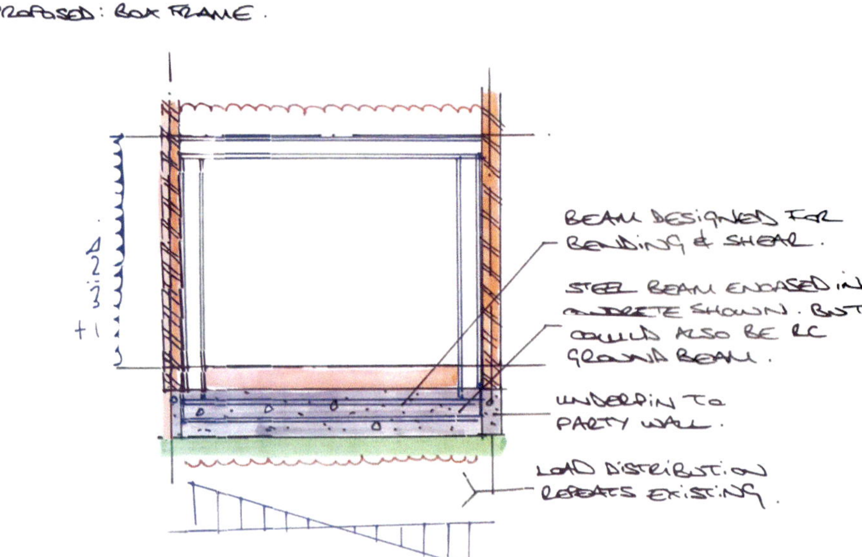

- Wholesale changes
  - More applicable to a large scheme where the buildings are empty. In this instance a suite of investigation and testing will dictate the extent of strengthening work required

Removal of walls in a masonry building impacts the innate robustness of a cellular masonry building. Impact on robustness always needs to be considered when changing the layout.

13.4.2 Removal of chimney breasts

Removal of chimney breasts in a building can be problematic, and specifically the use of gallows brackets to support the remaining masonry (Figure 13.9).

Current guidance on the topic can be found in Local Authority Building Control (LABC) advice[125] which many local authorities now rely upon. This gives comprehensive advice relating to when gallows brackets should be used, and what to consider when removing a chimney, which includes:

- Using gallows brackets only when the party wall supporting the bracket is a minimum 215mm thick, in brickwork, and in sound condition
- When using gallows brackets check that the neighbour's chimney breast on the other side of the party wall has not been removed (or partly removed)
- Bolts must be drilled into sound brickwork, not mortar joints
- The condition of the brickwork is critical, and there may be areas of the country where this option may not be acceptable to building control due to known problems
- The minimum height of the retained chimney breast below the roof line must be equal to or greater than the height of the brickwork above the roofline

These points should be considered when assessing existing gallows brackets encountered on-site. As gallows brackets rely on the withdrawal capacity of anchors and the inherent strength and condition of masonry for stability, there is a significant risk of this being jeopardised by the removal of brickwork during refurbishment works, if the use of gallows brackets on the opposite side of the wall is unknown.

For this reason, it is recommended that steel beams be adopted as the preferred solution to support a chimney when the breast is removed, as a safer immediate solution and to futureproof the building. Note that the removal of a chimney breast from a party wall is notifiable under the Party Wall etc. Act 1996 and will also reduce restraint to the overall wall length.

Figure 13.9: Gallows brackets

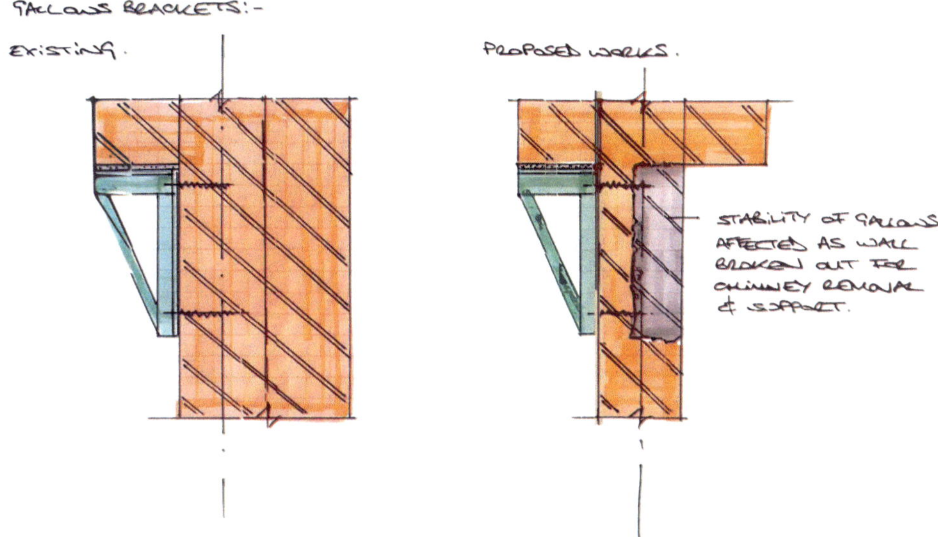

13.4.3 Historic strengthening in buildings

It is not uncommon to find historic strengthening works during investigations to the existing structure. It is not always obvious what these works relate to, with some structures having undergone multiple refurbishments during their lifetimes and some elements now redundant but left *in situ*.

It should be stressed, however, that the risk remains that a collapse could occur in the event of the strengthening elements being removed, either knowingly or unknowingly. In addition, consideration must be given to use of strengthening and/or replacement techniques, and knowing when to apply the appropriate option.

The decision-making process on when to strengthen or replace is discussed in The Concrete Society Technical Report 55[126].

When applying new strengthening works, strengthening should only be considered when the ultimate limit state (ULS) of the existing element is still greater than the proposed serviceability limit state (SLS) of the strengthened element.

For example, an existing timber Bressummer beam is proposed to be strengthened with two steel parallel flange channel (PFC) sections either side under a proposed new loading. In the event that the new PFCs are removed, the beam will fail under proposed SLS considerations, but will ultimately still be stable and robust under ULS design, preventing sudden collapse or failure.

Where the proposed SLS exceeds the existing ULS, replacement of the structural element should be adopted to mitigate the risk of collapse in the event of future works.

When altering an existing building the design engineer should be aware that strengthening work may have previously been undertaken.

13.4.4 Specific risks when undertaking alterations

13.4.4.1 Removing walls in a multi-storey, occupied building

Designing and constructing new structure to enable walls to be removed in multi-storey and occupied buildings should be approached with great caution.

Removing a wall at the top or bottom storey of the building has the practical benefit of the contractor being able to access the roof or foundation zones to carry out strengthening works.

However, providing structure for the intermediate storeys, while maintaining stability to structure above, and replicating the load paths so that the structure below is unaffected, is more challenging as access is limited by building occupation.

The temporary works solution for the intermediate storeys would be 'balanced needles' (a needle top and bottom of the floor with props at each end of the needle to balance the vertical load transferred from top to bottom), which then allows a box frame (picture frame) to be installed. The consequence of this is the formation of a step at floor level and a downstand, which an architect or client may not find acceptable.

The box frame solution is often changed to a portal frame or even a beam on padstones, which not only provides no lateral resistance for the wall section removed, but also changes the vertical load paths and redistributes the load to the structure below. This would not be acceptable under current UK law without detailed calculation. Paragraph 4(3) of the Building Regulations states:

"Building work shall be carried out so that, after it has been completed…it…is no more unsatisfactory in relation to that requirement than before the work was carried out."

If every tenant in the building were to remove a wall, then the altered structure would be a drastically different structure from the one initially constructed, with a much greater risk of collapse either due to greater stress in the loadbearing elements or much less innate robustness in the structure (Figure 13.10).

Buildings do not have an indefinite ability to absorb changes to the structure continuously across their life spans. At some point a limit is reached where the redundancy of the building is lost and the building is a shell of its former self. If the current building stock is to be retained, each structure deserves both care and attention at a local and global scale to ensure the structure maintains a safe level of robustness.

Figure 13.10: Apartment block with new openings

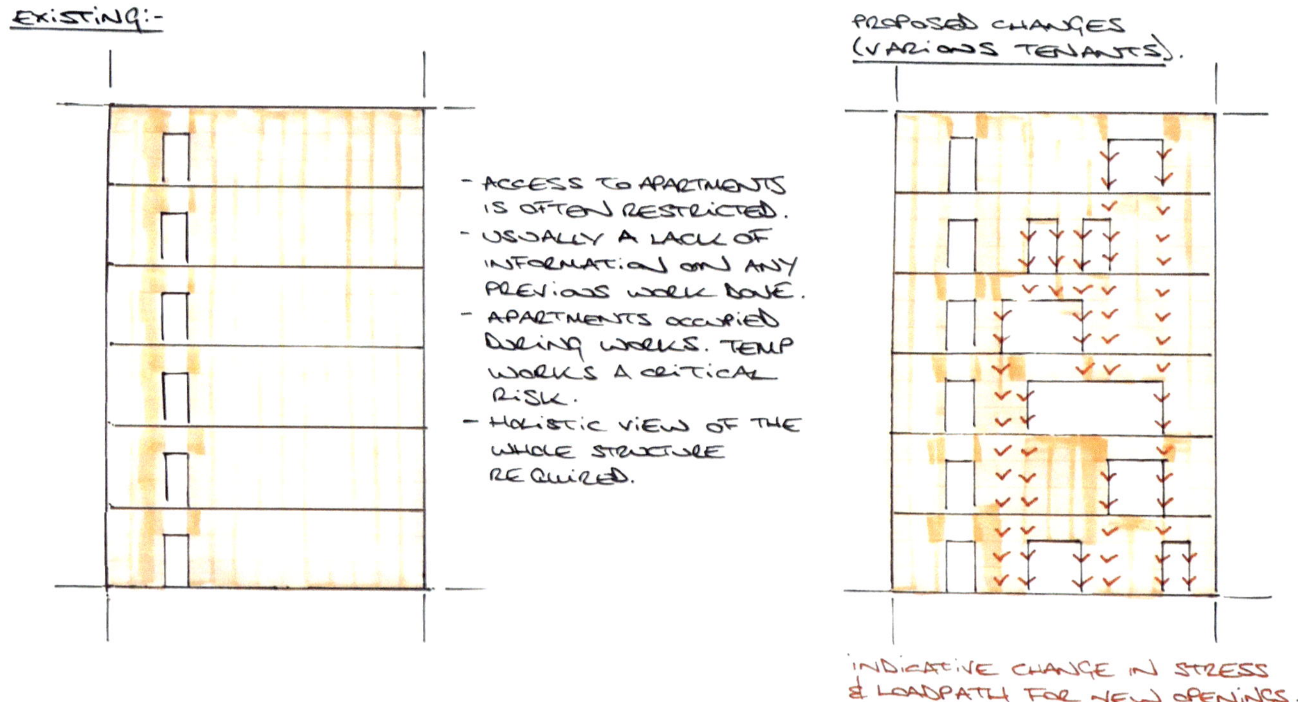

13.4.4.2 Eccentric foundations

Eccentric pad foundations are often seen when columns are pushed up to party walls in terraced housing on refurbishment projects (Figures 13.11 and 13.12). Justification of the foundation is often sought through load spread or the combined action of reinforcement within the pad footing. In reality the pad acts as a cantilever, without the rigid moment connection — as shown in Fig. 13.12 where the scheme has been flipped.

Figure 13.11: Eccentric foundations

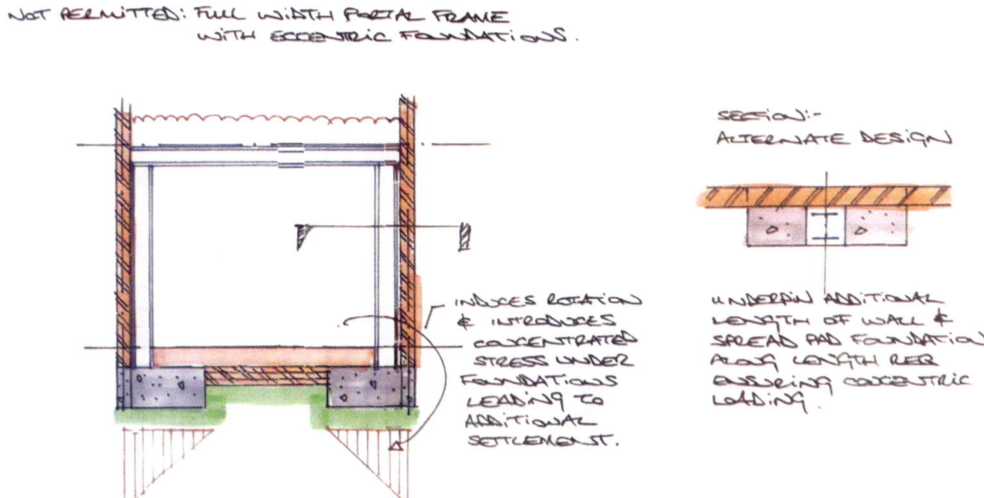

Figure 13.12: Eccentric pad foundations

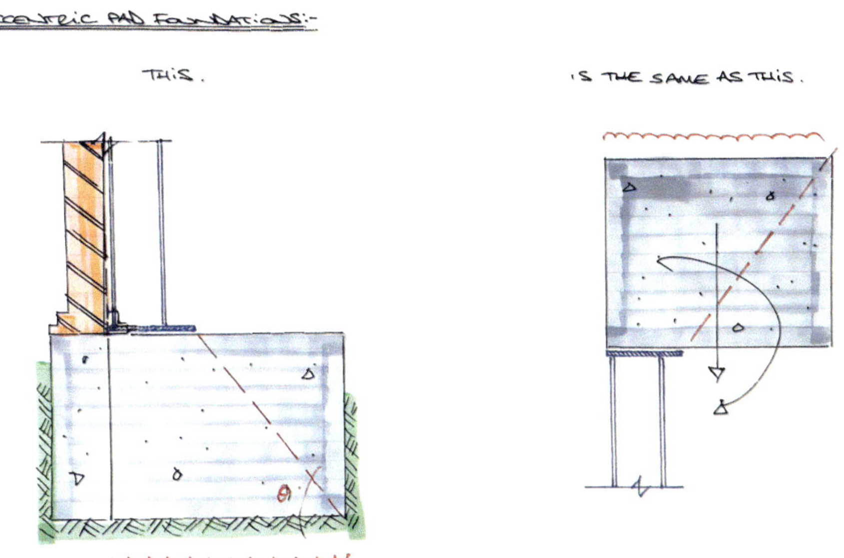

To ensure a uniform load spread, a moment connection would need to be incorporated between column and pad footing and the pad foundation designed as a reinforced beam to ensure all forces were resolved. Generally though, this moment is too high to justify and a nominal pinned base is used in practice. As a result, a degree of rotation is introduced into the pad foundation which can then create an overstress locally below the column.

With an eccentric foundation the pressure zone engaged under the foundation can be up to twice the permitted design load, creating a local overstress in the ground underneath the pad, which induces additional settlement and further impacts the structure above.

In practice, the rotation of the pad foundation will take restraint from the adjacent wall or underpin, introducing a lateral pressure on the structure previously not considered, as the system seeks to find equilibrium. It is not recommended to adopt eccentricity within structural components where this can be avoided.

13.5 Adding storeys to existing buildings (airspace development/vertical extensions)

Rooftop extensions have been undertaken for a number of years, but the industry as a whole is still very new, and the guidance available to engineers is limited. The recent change in permitted development to allow two additional storeys has prompted an increase in rooftop development schemes, and the formation of the Association of Rooftop & Airspace Development (ARAD)[127] seeks to address the lack of knowledge and understanding in this area. With good structural engineering involvement this is usually possible but may not always be economically viable.

The first distinction to make when assessing the feasibility of the additional storeys is whether the building is empty or occupied:

- With an unoccupied existing building there is full access for strengthening and building works. The whole scheme is essentially a refurbishment with a conversion/extension. It will be easy to undertake, subject to budget
- With an occupied building there is limited or almost no access for strengthening and building works. The whole scheme is reliant upon careful assessment of the existing structure, which is more difficult if access is limited

The design engineer needs to consider three aspects of building design:

1. Transfer of vertical loads to the ground, including the extra loading associated with the additional storeys.
2. Transfer of extra wind loading to the ground.
3. Robustness and disproportionate collapse.

In an ideal situation, where the existing building is empty and fully-accessible the overall process should be as follows:

1. Visual condition survey.
2. Investigation and testing.
3. Feasibility report, which includes but is not limited to:
   a. Disproportionate collapse strategy
   b. Review of the scheme and identification of key risks
   c. Quick stress check (less than 10% increase)

However, where the building is occupied it is prudent to concentrate on steps 3a, b and c before a fee proposal is even submitted, so the process becomes:

a. Disproportionate collapse strategy
b. Review of the scheme and identification of key risks
c. Quick stress check

Followed by:

1. Visual condition survey.
2. Investigation and testing.
3. Feasibility report.

13.5.1 Feasibility report for a vertical extension

A feasibility study is an exercise to assess the design and cost implications of a proposed project; the end product is a feasibility report. It may contain photographs, sketches and drawings as well as financial projections.

It is not, however, an exercise in listing all the potential problems with the scheme and forcing those further down the design process to be dealt with later. For example:

BAD. The walls are overstressed and should be tested to confirm feasibility.

GOOD. The walls are overstressed, but the proposed stress within the wall still falls under the minimum permissible load of $0.42N/mm^2$ based on a 100mm-thick wall. Therefore, walls can sustain the increase in proposed load (i.e., the scheme is feasible).

A good feasibility report should be able to detail the next steps for the non-technical reader, as well as addressing the technical points that only an engineer will understand and appreciate.

The feasibility report should:

- Contain a CDM regulation assessment regarding construction method — whether the build sequence has been considered, if is it a tight site with limited access to cranage and deliveries, etc.
- Provide evidence that the proposed scheme is achievable (i.e., stress and strength checks), for which supporting calculations will be required
- Confirm disproportionate collapse strategy, including understanding the structural performance of the building under accidental loading. This is also likely to include a risk assessment with proposed mitigation measures
- Eliminate any risks to the existing structure showing extent of strengthening works to be carried out where applicable (the design of which can be included within stage 4 but specialist strengthening work elements should have been approached and confirmed as achievable as part of the feasibility report stage)
- Provide sketch/plan of required transfer structure (grid/column layout) between proposed and existing structure. This should be taken to stage 3 (member sizing) which allows the developer to undertake a costing exercise of the required works prior to adding the new-build element of the design

The feasibility study then draws a neat line in the design process should the client wish to tender the contract as a design and build submission for the new-build portion of the works, or go down a traditional procurement route. If the former, the principal engineer should be retained to ensure that the proposed works are detailed and constructed properly.

Architects are typically the first point of contact for any proposed development. However, on large schemes, and specifically those where the existing structure is occupied by various leaseholders, it is recommended that the initial design is led by the structural engineer (appointed pre-planning). Equally, a building that may seem appropriate for the structural engineer may not work for the fire engineer, so ideally a multi-disciplinary team should be involved from the start.

A rooftop extension will be dictated by the strength, condition and layout of the existing structure below. A lot of time, money and effort can be wasted early on in drawing up a scheme which, without engineering input, may not include the ability to incorporate strengthening works where required, and ultimately leads to wasted time and thwarted expectations. Common issues which may stop a scheme before it has even started are:

- The existing structure is formed using a lightweight frame (timber stud or lightweight steel frame). The lightweight construction cannot sustain the increase in vertical and lateral loading. As modern methods of construction, these structures have been value-engineered for optimal performance under current design codes, and do not contain the same amount of 'spare capacity' as some older structures. The same argument can be applied to some modern steel-framed structures with very slender columns and bracing
- There is an existing transfer structure within the lower floors (beam, slab, floor) which may be overstressed at a lower floor with no access for strengthening. Typical examples include inset storeys or elements within clear-span open commercial layouts
- Internal foundations and structural elements may be overstressed with no access for strengthening
- Large existing openings may be present with no access for inspection or strengthening, e.g., a row of terraced commercial shop fronts
- The condition of the building may be poor, e.g., cracks, subsidence, local areas overstressed, general condition of the materials, poor workmanship or construction (generally picked up within the visual condition survey)
- Planning restrictions — when no allowance has been made for a new transfer structure (approx. 1m in depth) within the planning drawings which would result in an overall increase in building height required. While this would not necessarily halt a scheme, the impact on the overall duration of the project should planning need to be resubmitted can be considerable

Many schemes may not make it past the structural engineer's quick assessment of key risks. Beware of schemes that have achieved planning but without a structural engineer's input. Vertical extensions are perhaps the only example in the construction industry where construction sites are intentionally placed above a live residential or commercial unit.

The risk to residents both short-term and long-term cannot be stressed enough. Not every building is suitable for a vertical extension and it is perfectly acceptable to walk away from projects where they will not work — many structures simply are not suitable for vertical extensions. The structural engineer is not only taking responsibility for the proposed structure, but also the existing structure — including the safety and wellbeing of residents. These are very high-risk projects and should be treated as such.

13.5.2 Increase in vertical loading

It may be very difficult to justify the increase in load when little is known about the existing structure. Often, access to the existing building is restricted, especially if a building has multiple leaseholders or tenants.

The top two storeys of any vertical extension will always be overstressed, as these are the lowest stressed elements within the current structure. Moving down the structure it is important to review the percentage of overstress in relation to any transfer structures identified. Figure 13.13 shows that within the +1 storey scheme:

- The first-floor podium will be approx. 29% overstressed
- Ground-floor walls within the loadbearing masonry section are 23% overstressed
- The foundations in both options are approx. 19–20% overstressed

These overstress values contain an allowance for the vertical component of increased lateral loading. In this context 'overstress' means the additional stress that the existing building is experiencing, not that the elements are overstressed beyond structural capacity.

Figure 13.13: Quick overstress assessment

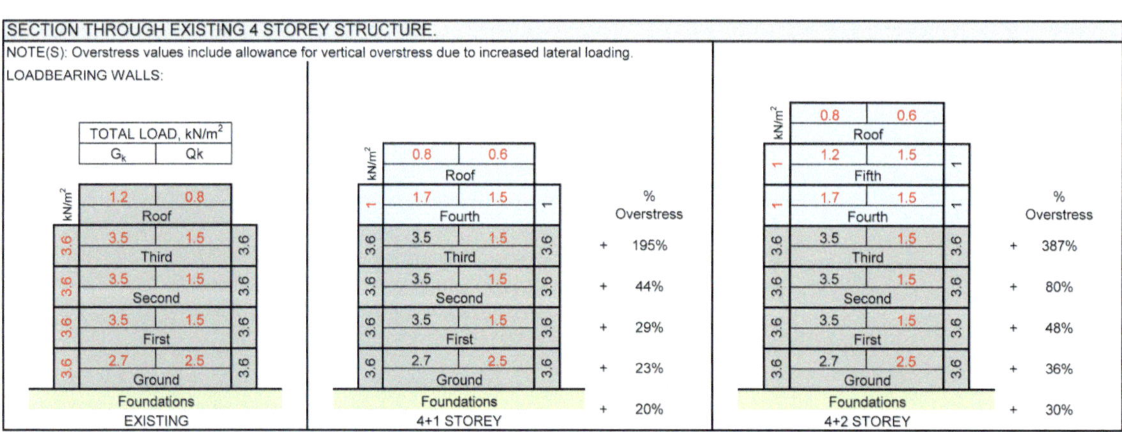

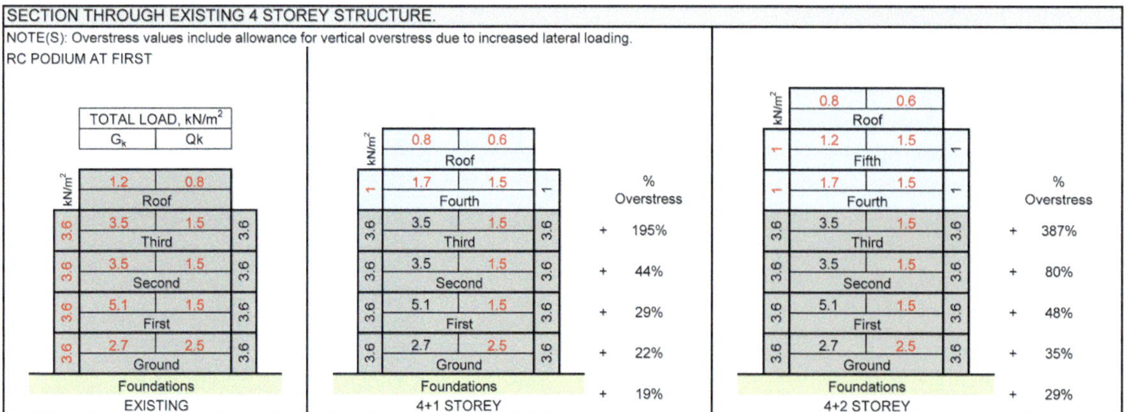

The priority should be to identify areas of high overstress where access is restricted or not available. Areas of high overstress can still be reviewed and investigated, but without any previous construction information (e.g., drawings or calculations) justification for the overstress cannot be made and this quick exercise may stop a scheme before it has started.

Where access or information is sufficient to progress the design, areas for investigation and testing should be drawn up together with a more detailed vertical load take-down, and a strength check on the accessible areas targeted for investigation.

The aim of the load take-down should be to verify any expected results from investigation and testing. It also derives values to be used for the strength check. The strength check should be used to determine the capacity of existing building elements (beams/walls/columns/foundations) and verify their strength under current and proposed vertical loading, and the vertical component of lateral loading. This ultimately allows an assessment to be made of any redundancy contained within each element, and the feasibility of any future vertical extensions above.

Walls subject to a 44% overstress at second floor, for example, were found to be, through investigation and testing and a literature review of existing calculations/drawings, acting at 53% capacity of their strength. Increasing the stress by 44% brought the wall up to 76% unity, confirming that the wall has capacity for an additional storey to be applied.

13.5.3 Increase in lateral loading
While lateral loading is an important design consideration, some forms of construction are impacted by an increase in lateral loading to a greater extent than others, and its effect on structure will be largely down to the form of construction of the existing building.

Figures 13.14–13.17 offer an overview of an increase in lateral load acting on foundations.

Figure 13.14: Increase in lateral loading

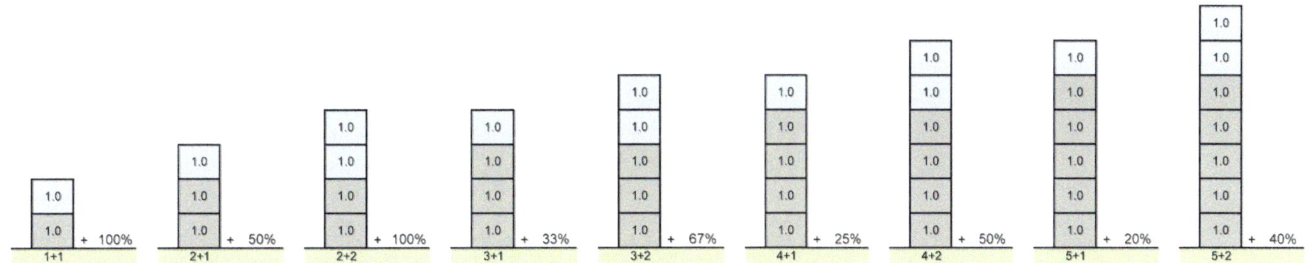

The magnitude of the increase in lateral loads due to height is in proportion to the height of the building. However, the increase in the vertical loads on the structure due to lateral load is in proportion to the height squared. As an example, if an existing three-storey building has two storeys added, making it a five-storey building, there will be an extra 67% of lateral loading (or 1.67 x existing lateral load), which could translate to vertical loads on columns of up to $1.67^2 - 2.8$ x original vertical load due to lateral loads.

$$(5 \div 3) - 1 = 0.67 = +67\% \text{ increase in lateral loading}$$

Increased lateral loads will impact vertical loadbearing members, any bracing members, connections in moment frames and other structural elements (Figs. 13.15–13.17).

Figure 13.15: Lateral stability review: loadbearing walls

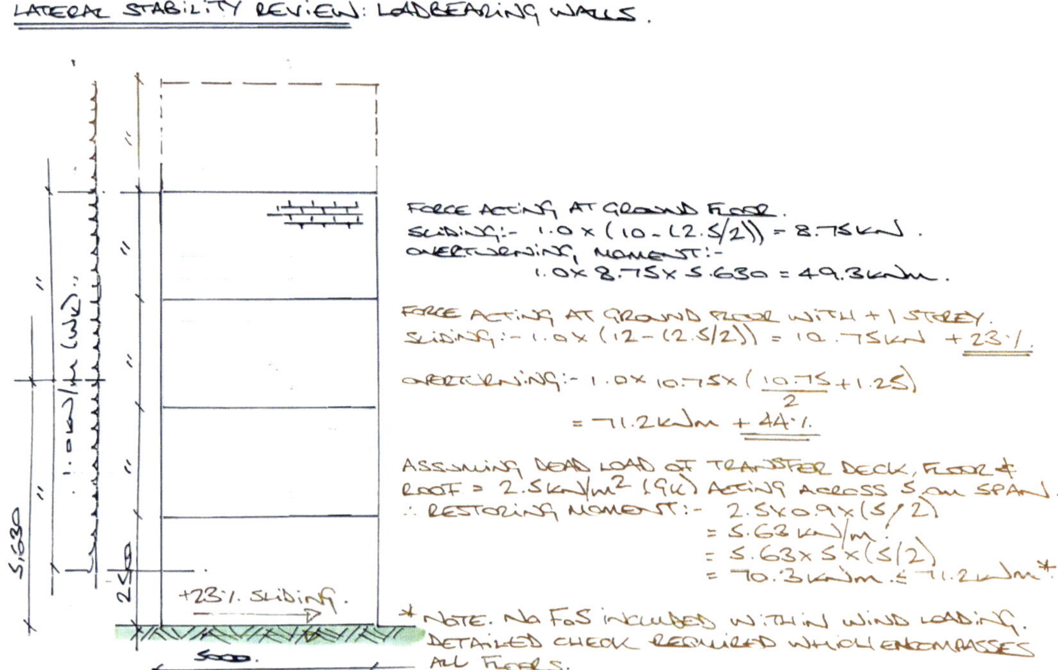

Figure 13.16: Lateral stability review: braced frames

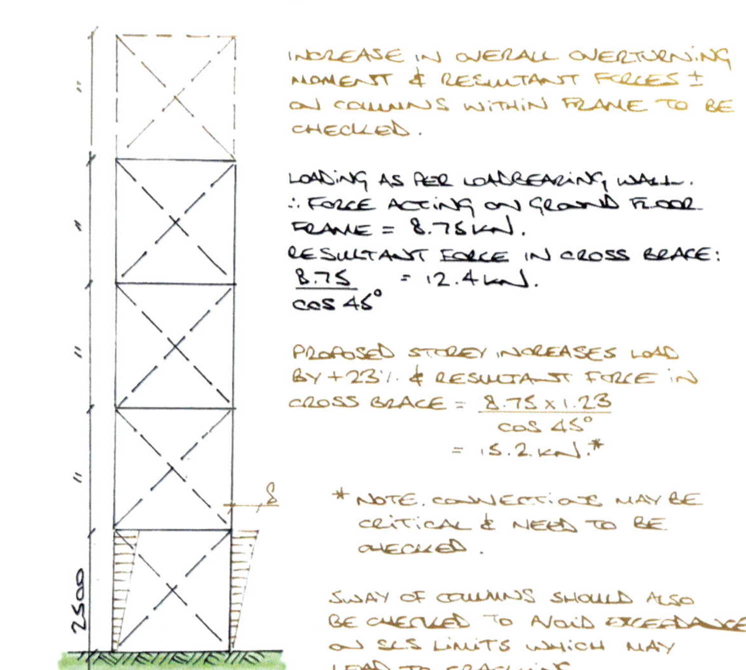

Figure 13.17: Lateral stability review: moment frames

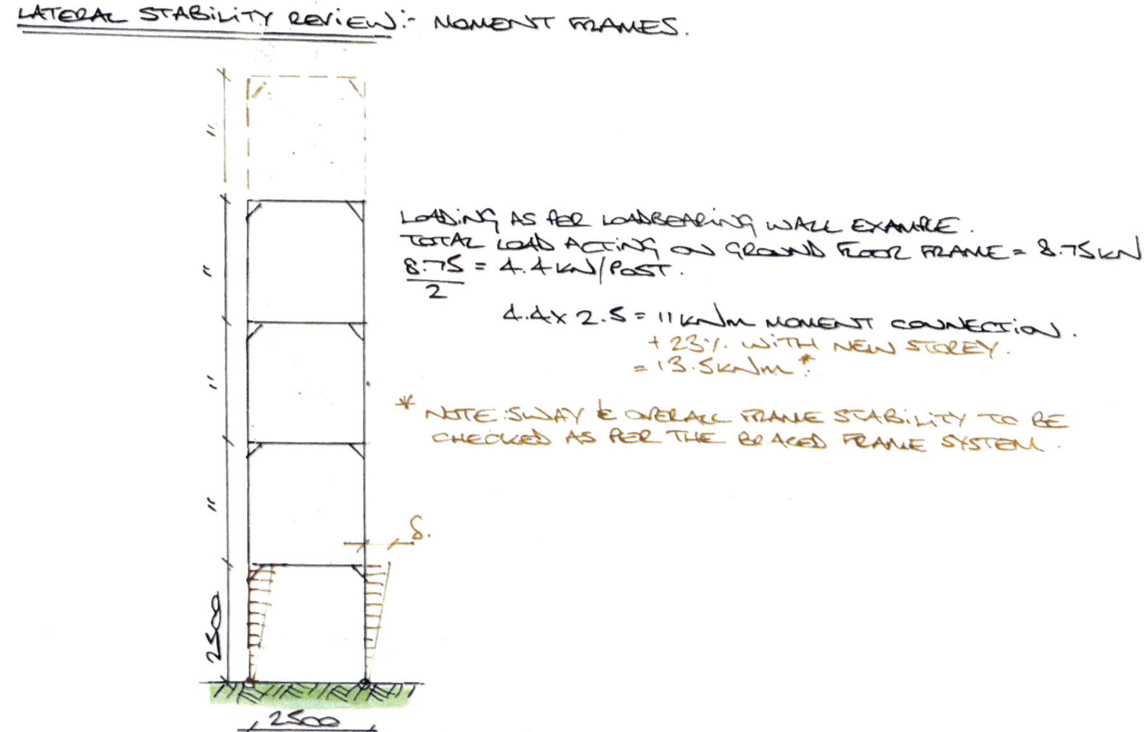

13.5.4 Disproportionate collapse considerations
Compliance is often sought when the proposed scheme changes the consequence class of the building (e.g., from CC2a to CC2b with regard to the recommendations made in Approved Document A) or when building work is triggered as a result of a material change of use in accordance with the Building Regulations.

As an example, if a residential building is currently four storeys and an extra storey is added, the consequence class would change from CC2a to CC2b. It is unlikely that the existing building will have been designed to the requirements of a CC2b building and, therefore, the structure of the original four storeys will need to be reviewed with respect to disproportionate collapse; a requirement triggered by the building work definition in the current Building Regulations.

There is a further complication — before 1970, robustness and disproportionate collapse were not considered in building design. It was the collapse of Ronan Point that triggered a review of the building regulations and led to the inclusion of robustness. Therefore, any building built before 1970 will not meet current requirements, even before the addition of an extra storey.

13.5.4.1 Past guidance: the Camden Ruling
The Camden Ruling[128] is a version of strong floor design. The Department for Communities and Local Government (DCLG) report on robustness[45] is a great starting point for engineers designing vertical extensions and the publication includes a discussion on the Camden Ruling.

The Camden Ruling was an internal document produced by the London District Surveyors' Association (LDSA), and was available for distribution among local authorities. The theory was that the new storey was constructed such that any damage occurring within the new storey would be contained by the new floor forming the extension, or the roof of the existing storey below. Adhering to this the alteration would not change the risk to occupants below, and therefore the original structure below could be assessed as CC2a as it was before.

The original Camden Ruling document appears to be lost, but it is referred to in the first edition of this *Guidance* and also in the DCLG report which states:

"The Camden Ruling is controversial because the construction of additional storeys is almost certain to increase the risk to the occupants of the lower floors. Consequently an approach based on compliance so far as reasonably practicable with the current regulations as is the case in Scotland is preferred."

The DCLG report is not in favour of the Camden Ruling although the reason is not explained, other than that it increases the risk to existing tenants below, which is contrary to what the ruling implies. It may also be because of the increase in load on the existing structure.

The aim of the Camden Ruling was to carry the debris from the collapse of any new structure above onto a strong floor, and protect the existing tenants below. However, to achieve this the floor first needs to be resistant to an accidental load ($34kN/m^2$) in order to be able to support any debris load thereafter.

Designing to this requirement would result in a concrete slab, approximately 250–300mm thick, depending on the spans involved. The weight of that slab is equivalent to approx. two storeys of new lightweight structure. Therefore, the new floor adds significant weight to the existing building below, in addition to the new storeys themselves. This adds additional risk to the building and its occupants on the lower floors. The Camden Ruling does not consider what would happen if there were a collapse in the lower, original floors; if that were to occur, the weight of the resulting falling debris would be considerably greater than before the addition of the new storeys.

Therefore, with regard to proposed vertical extensions, it should be considered if the aim is to:

- Design to protect the existing structure below from potential collapse of the new extension above (the Camden Ruling), or
- Prevent both the new and existing structure from collapsing if there is an accidental event

It is felt that it is the latter approach that should be adopted going forward.

Due to the age of the now lost Camden Ruling, the increased frequency and scale of vertical extensions and the discomfort expressed within the DCLG report, the use of the Camden Ruling is not acceptable to justify vertical extensions and a risk-based approach should be adopted.

13.5.4.2 Using a risk-based approach

Guidance suggests that a review of a building starts with the premise that the building *"is no more unsatisfactory in relation to that requirement than before the work was carried out"* once the work has been completed. If the additional storey changes the consequence class of the building, the building also must be compliant with the Building Regulations throughout its height (Fig. 13.1). Taking a risk-based approach for alterations to existing buildings is discussed in the IStructE's *Manual for the systematic assessment of high-risk structures against disproportionate collapse*[19] and may offer the safest approach when looking at constructing additional storeys on an existing building.

Typical risks include:

- The additional storey could overstress the elements below, both vertically and laterally
- Access to the building may be restricted for strengthening works, leaving the design engineer uncertain about the existing structure
- A change in storey height may trigger a change in disproportionate collapse classification
- The consequence of an accidental event may be greater due to the increased building height and weight, but the chance of an accident occurring in each storey (as well as in the new proposed storey) remains the same
- Open-plan layouts could mean there is a large-span area that is different from the floors above, and could have a detrimental effect on higher storeys, which increase the consequences if an accidental event occurs
- Errors in the original building design

Structure-specific risks should also be identified.

As an example, a 2010 residential building comprises two distinct structural forms. Figure 13.18 shows a four-storey masonry building of loadbearing construction with precast floors at either end of the U-shaped building. There is also a three-storey masonry building over a concrete podium at first floor which provides car parking to the undercroft below. The shallow roof profile is formed using lightweight trusses.

Figure 13.18: Structure-specific risks identified

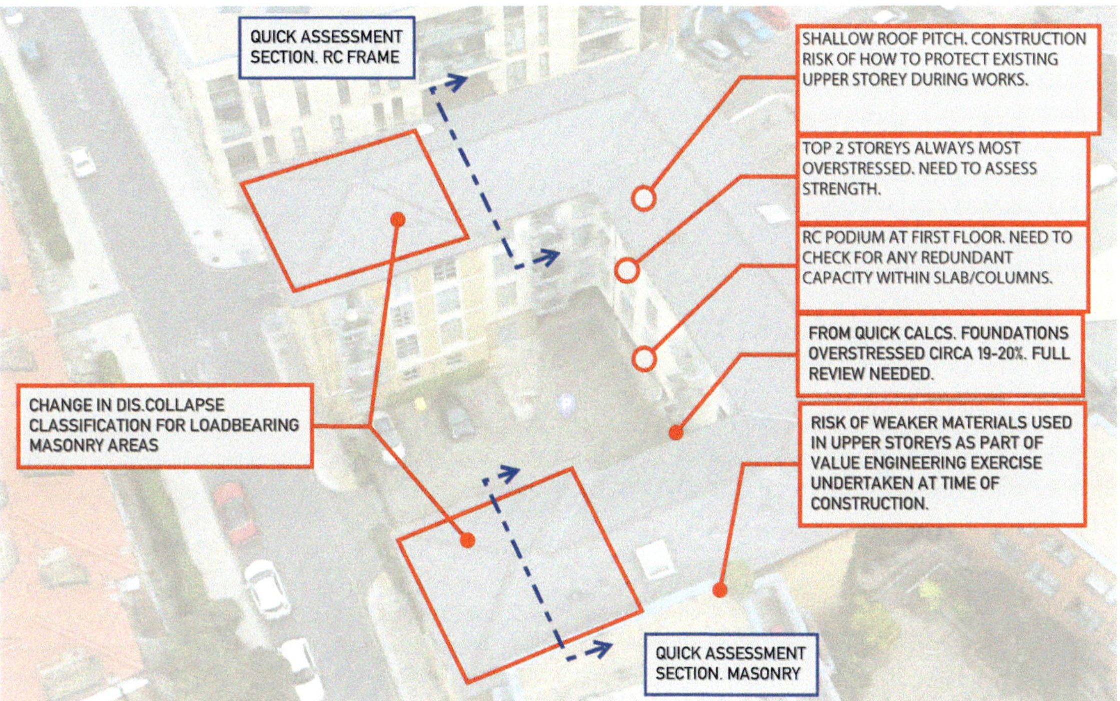

As the building is occupied, a quick stress check should be conducted as part of the specific risk assessment, to determine whether the scheme is feasible (Figure 13.19).

Figure 13.19: Quick stress check

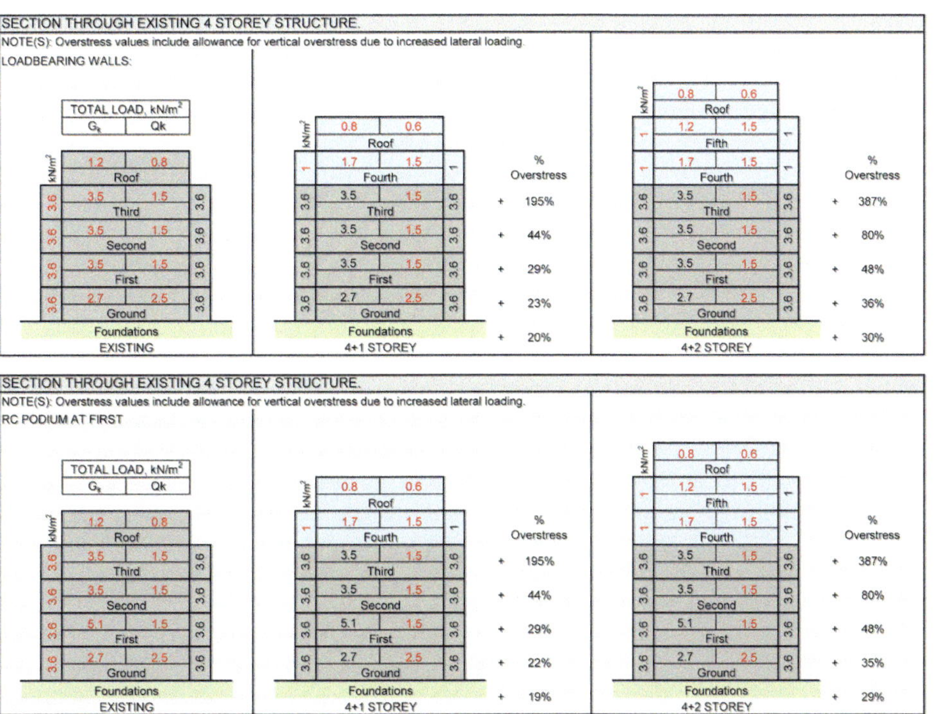

Working through the risks

As this building is a 2010 structure, the health and safety file should be available which will contain the original structural engineering calculations and specification for the building, helping to mitigate many of the risks. Both a +1 storey and a +2 storey option have been considered within the quick stress check. However, only the +1 option is considered going forward due to the magnitude of the increase in stress.

Vertical loading overstress

In this instance:

- The first-floor podium will be approx. 29% overstressed
- Ground-floor walls within the loadbearing masonry section are 23% overstressed
- The foundations in both options are approx. 19–20% overstressed

The values obtained are greater than typical values adopted for permissible overstress. Therefore the next steps are either a document review of the existing building control package (where applicable), or engaging in a suite of investigation and testing (where accessible).

A review of the original calculations would indicate that the masonry and piled foundations contained sufficient spare capacity to accommodate the increase in loads.

Access for investigation and testing

- In this instance a reduced scope of investigation works has been used to reiterate the findings (i.e., mechanical removal of concrete cover to first-floor slab and ground-floor columns to confirm reinforced concrete content as designed)
- Access is only available to communal areas, so investigation works are to be targeted within those areas
- An independent load take-down across each of the test locations helps verify the work of the original engineer, and also the *in situ* load/design

A change in storey height triggers a disproportionate collapse reclassification

If this occurs when generating the design, the following situations should be considered:

- The current structure of both the four-storey masonry building and the three-storey masonry over the podium are known to have been designed as CC2a throughout, which may include either horizontal ties (unlikely) or effective anchorage of floors to walls
- The four-storey masonry structure is reclassified as CC2b under the proposed scheme and needs to be checked for compliance
- The structure above the reinforced concrete podium and the reinforced concrete ground-floor structure should be checked as a CC2b structure in its entirety
- Statute requires that buildings are resistant to disproportionate collapse. Approved Document A provides guidance on how to achieve that in the form of vertical and horizontal ties, notional element removal and key element design. Knowledge of these approaches and specific approaches to steel, concrete, masonry and timber disproportionate collapse design is required when dealing with existing buildings, as the proposed strategy may vary depending on the form of construction of each existing structure
- As the existing structure is fully-occupied with no access for vertical and horizontal ties to be provided, the revised building will need to be checked with reference to the Building Regulations and be 'no more unsatisfactory than it was before' through a risk assessment approach. The specific risks are vertical load and horizontal load overstress and disproportionate collapse
- In addition, because of the change in consequence class and because the building was compliant with the Building Regulations before the alterations, the whole building must still be compliant with the Building Regulations once the works have been carried out
- The risk of an accidental event to the existing structure on the lower floors remains the same, but the consequence may be higher with the addition of the extra floor above. As these are loadbearing masonry walls, a $2.25h$ notional length of wall should be removed across all areas of the building and the resultant floor areas that collapse assessed under permitted limits stated in Approved Document A, with a check of the masonry arching taking effect above. Other consequences should also be assessed

- The contribution of the structural topping on the precast floors is also considered to see if collapse is not just limited but preventable. It is demonstrated through calculation that if an external wall supporting the floor is blown out, the structural topping in the floor would enable the floor to span in the opposite direction. However, the reinforced structural topping would not necessarily be able to carry the weight of the external wall over (Case study 11). It was concluded that this consequence, with or without the additional storey, remains the same because of the new steelwork ring beams at third-floor level
- Where an accidental event/explosion might take place in the existing third floor, a steel ring beam will be placed across the top of each loadbearing wall to ensure that the proposed structure above can span the required $2.25h$ of missing wall and avoid collapse
- Where an accidental event/explosion might occur in the new upper storey, the Eurocode must be referred to. BS EN 1991-1-7, Clause 5.3(4) states:

"The explosive pressure should be assumed to act effectively simultaneously on all of the bounding surfaces of the enclosure in which the explosion occurs."

- The largest room in the proposed scheme is assessed on the assumption that if the floor blows out, the amount of structure that collapses is still below the permitted recommendations stated in Approved Document A
- An explosion may occur in the new upper storey, resulting in the floor blowing out and a debris load being applied to the existing (old) roof below. It was concluded that in this sequence of events, coupled with a vertical and lateral design check of the existing structure, the risk of an explosion and consequences of that action, with or without the additional storey, remain the same, as the provision of new steelwork ring beams forming the transfer deck would not increase the consequence of accidental events. Therefore, 'the building is no more unsatisfactory than before'.
- The engineer has carried out the appropriate design checks to understand the actual structural performance under permanent, imposed and accidental loading
- A detailed risk assessment has been undertaken and the engineer is satisfied that risk to the inhabitants of the building has not increased by adding an additional storey

These risks could be added into a schedule (Figure 13.20). For a greater understanding of the likelihood and consequence of actions relating to risk and the impact when considering explosions within a structure, refer to the IStructE's *Manual for the systematic risk assessment of high-risk structures against disproportionate collapse*.

Figure 13.20: Hazard identification and mitigation schedule

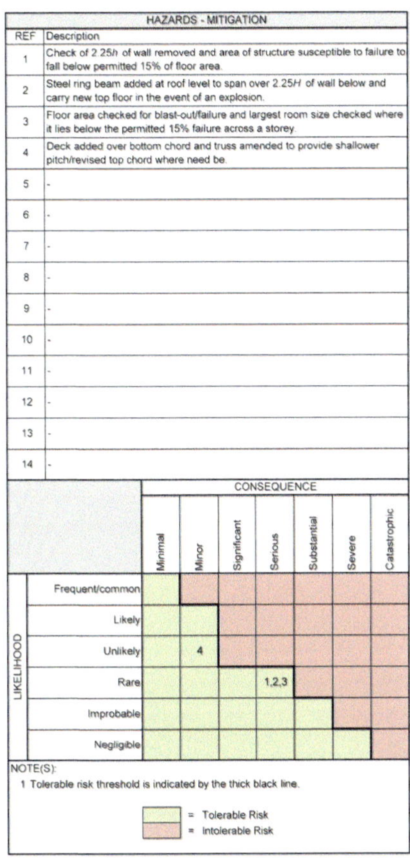

HAZARDS - INTIAL ASSESSMENT	
REF	Description
1	Explosion to ground, first and second floor of existing building.
2	Explosion to existing top storey.
3	Explosion to proposed new top storey.
4	Construction hazard. Crash deck of safe method of working to residents of existing top floor as roof removed.
5	-
6	-
7	-
8	-
9	-
10	-
11	-
12	-
13	-
14	-

HAZARDS - MITIGATION	
REF	Description
1	Check of $2.25h$ of wall removed and area of structure susceptible to failure to fall below permitted 15% of floor area.
2	Steel ring beam added at roof level to span over $2.25H$ of wall below and carry new top floor in the event of an explosion.
3	Floor area checked for blast-out/failure and largest room size checked where it lies below the permitted 15% failure across a storey.
4	Deck added over bottom chord and truss amended to provide shallower pitch/revised top chord where need be.
5	-
6	-
7	-
8	-
9	-
10	-
11	-
12	-
13	-
14	-

Provided each of the risks has been identified and resolved, the new scheme can be considered feasible.

Ultimately, it is believed that the approach of using notional element removal should be applicable for all masonry buildings which house three or more apartments. Anything smaller than this should be examined on a case-by-case basis, as the largest room in a single dwelling is likely to exceed 15% of the entire storey area and in the event of an explosion the floor area at risk of collapse would be greater than permitted limits.

It should be stressed again that this is not a prescribed, universal method. It is a risk-based approach that must be applied systematically to each building and storey. It is not always possible to retrofit ties into old buildings, and that act itself could make the building less robust in some instances, but those risks need to be demonstrated clearly to the building control authority reviewing the scheme, so they have a comprehensive understanding of the hazards and consequences.

Some buildings contain ties which can be extended, but some schemes may not have the resources (time, money, practicality) to retrofit ties into existing buildings. However, there are other safety measures that can be adopted such as additional (fire fighting) lifts, upgrade of communal stair cores, removal of hazards from within the building, etc. Each of these actions helps to reduce the overall risk to the building — all of which can be identified in a risk assessment of the structure.

A detailed understanding of the existing structure and its risks requires a high level of engineering competence. It must be communicated in a clear manner, as not all professionals reviewing the scheme (including those in building control) will have the same level of technical understanding of building structures as the design engineer.

The end-users are important here, since they will need to bear these risks. Structural engineers have a professional responsibility to look after building occupants to the best of their ability and should be prepared to justify the proposed design (and safety measures) adopted or omitted whenever required.

13.5.4.3 Debris loading

There are no references to debris loading in Approved Document A or within the Eurocodes.

The Department for Communities and Local Government report references the London Building (Constructional) Amending By-Laws 1970[129] which states that the debris load could be taken as a static load equivalent to its own weight, plus the reduced imposed load. However, if the floor consisted of precast units not tied together over their supports (i.e., simply-supported), the debris load should be taken as three times the weight plus a reduced imposed load.

This is probably a sensible way of assessing debris loading but, of course, if there is anything unusual about the building, it may be too simplistic an approach.

13.5.4.4 Proposed layouts and transfer decks

The main purpose of the transfer deck is to ensure the proposed vertical load paths are transferred to the existing vertical load paths below. The easiest way to achieve this is to align proposed loadbearing elements in the new extension with those in the structure below. Failure to do this will lead to a material increase within the transfer deck, as additional members are required to support the new elements and transfer that load back down to the existing structure.

Where the existing building is a framed structure, the proposed column grid should align with that below, and a new floor structure should be added between the columns where the existing roof is not suitable for reuse as a floor element. This will also aid with extending vertical ties up from the existing structure and into the new structure.

Failure to achieve this alignment will overstress potentially weak elements within the existing structure such as lintels, edge beams and slender piers, which may not have the ability to carry the additional load of the vertical extension. The reuse of existing flat slabs forming the roof of an existing structure carries an obvious advantage by avoiding the need for a new floor deck, and many planning applications submitted without input from an engineer may assume their retention as a floor within the proposed scheme. However, inclusion as a floor also comes with risks, the obvious one being loading (Figure 13.21).

Figure 13.21: Reuse of existing flat slab to form the roof of an existing structure

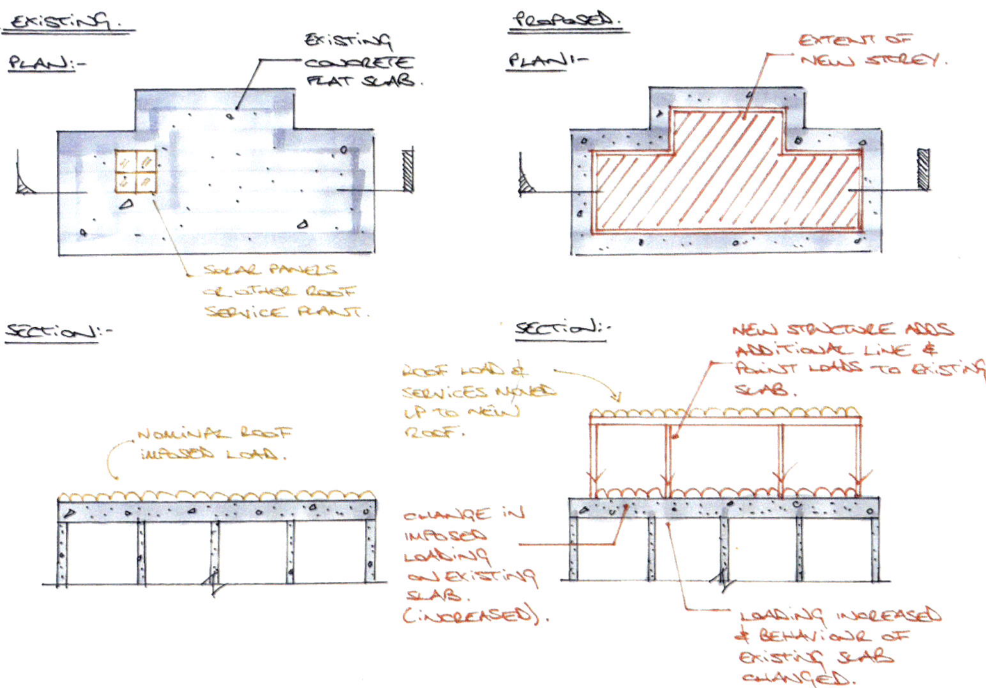

Many roof structures, old and new, have only been designed to carry a nominal roof-imposed load for access and maintenance purposes. A change of use generally sees the requirement for imposed loading to be doubled for a residential application (more for other uses), in addition to the roof load being retained but at a higher level.

Most rooftop projects are situated within built-up urban areas, requiring a smaller footprint than the structure below as the additional floor is set back to account for right-of-light requirements. This set-back introduces additional line and point loads on the existing slab where the new structure is placed upon it. The engineer should ensure that the existing slab contains sufficient strength and stiffness for the proposed ULS and SLS loading requirements, in addition to any punching shear checks around existing column heads below (where applicable).

This is a high risk as newer structures are value-engineered at roof level, while older roof structures may contain material which is a third of the strength of that used today. The engineer is advised to conduct a full design check on the existing slab, which will also need to include an accidental load check.

13.6 Existing masonry buildings — specific considerations

Masonry structures account for a large proportion of building stock within the UK and across the world. This type of building has been resilient to social change with single-occupancy large houses changed into flats, and homes with limited kitchen and bathroom provision altered to enable large open-plan kitchens and modern bathrooms.

As discussed in Section 13.2.1, some alterations to buildings will require the design engineer to check that the existing building complies with A3 of the Building Regulations.

In the built environment, it is not possible to incorporate a steel frame into every existing building to make it compliant with CC2b requirements for disproportionate collapse. It is inefficient and negates the current robustness within the existing building stock.

Incorporating other methods of compliance suggested within Approved Document A, such as notional removal of an element and a knowledge of each of the different approaches, allows a more practical approach to be undertaken.

13.6.1 Impact of layout and floor construction

The level of difficulty in designing for existing masonry buildings depends on the building layout and its occupancy. A cellular masonry building that has had minimal interference over its lifespan, is a more robust building and less risky to work on than a masonry building that has been repeatedly altered, or a masonry building with large open-plan spaces. A masonry building with an *in situ* concrete floor is likely to be more robust than one with a precast or beam and pot floor or timber floor.

13.6.2 Notional element removal approach for existing masonry buildings

If the engineer decides to take the notional element removal approach in checking an existing building for compliance, it will be necessary to ensure that the masonry walls are restrained at suitable intervals, because the length of wall to be removed is defined as the length between vertical restraints or 2.25h.

The requirement of 150kg/m² self-weight for partitions that are used for vertical restraint[130], rules out the use of lightweight structural partitions such as timber and lightweight steel frame acting as restraint for masonry walls.

When determining whether a building can withstand a wall being removed, calculating the floor area that will collapse provides a view on which buildings may require a steel-framed intervention. If a storey height of 2.4m is assumed, 2.25h = 5.4m. If the average floor span is 5m, the collapse area would be: 5.0 x 5.4 = 27m². If only 15% (or 100m² max.) is allowed to collapse, the approximate minimum floor area where a partial collapse could be tolerated is (27m² ÷ 15) x 100 = 180m². A smaller floor area is likely to need steel intervention or other forms of strengthening and tying depending on the type of structure, spans and structural arrangement involved.

When presenting a scheme to a client and to building control, the design engineer should set out clearly the strategy for disproportionate collapse, whether it is providing ties, looking at element removal and/or taking a risk-based approach to determine whether the risk of collapse has not increased. When in doubt the approach should be confirmed with building control at the earliest opportunity.

13.6.3 Arching of masonry

Much has been said about the ability of masonry to arch if an existing wall collapses, either due to an accident or when using the notional element removal approach in checking for compliance with A3 of the Building Regulations.

Calculations for checking arches such as those shown in Figures 13.22 and 13.23 generally follow the techniques explored by Jacques Heyman[131], ensuring that the line of thrust in the arch follows the shape of the arch.

Worked examples for designing arches can be found in the *Structural masonry designers' manual*[84].

Figures 13.22 and 13.23: Arches in masonry structures

However, in the case of a masonry wall being blown out, the masonry above is not jointed like an arch. The joints are horizontal and vertical rather than perpendicular to the arch that may form.

It can be seen in existing buildings that, when a beam has deflected too much, masonry does indeed arch, even with horizontal and vertical joints (Figures 13.24 and 13.25).

Figures 13.24 and 13.25: Masonry arching over deflected lintels

The design engineer could explore how the thrust in the arch might translate into vertical and horizontal forces across the joint (Figures 13.26–13.28) and check that the masonry and mortar joints have sufficient capacity to resist these forces.

Figure 13.26: An arch forming over a 'missing' wall

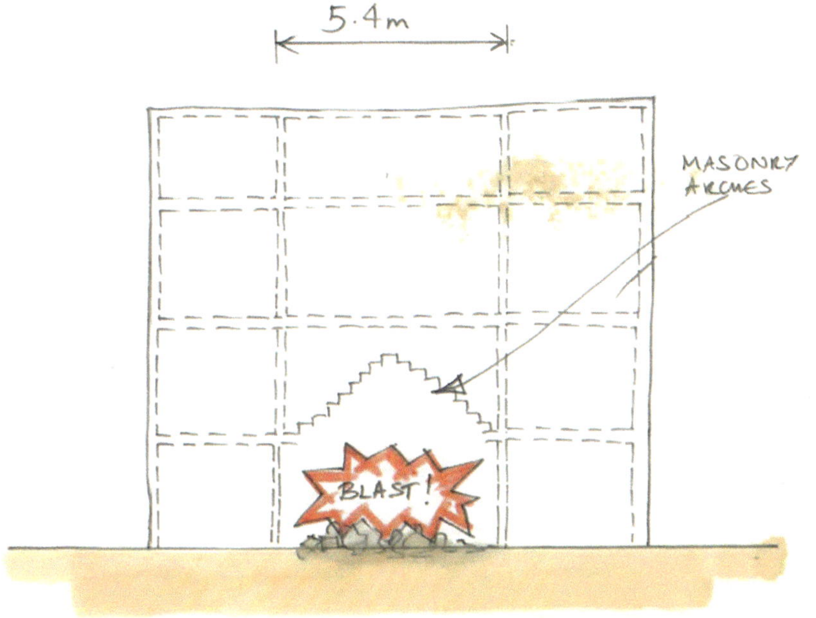

Figure 13.27: Approximate calculation of forces within the arch

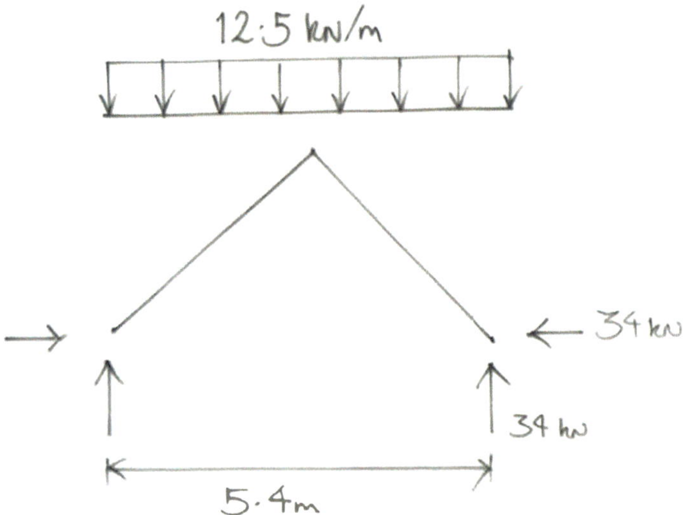

Figure 13.28: Resolving arch forces into the joints

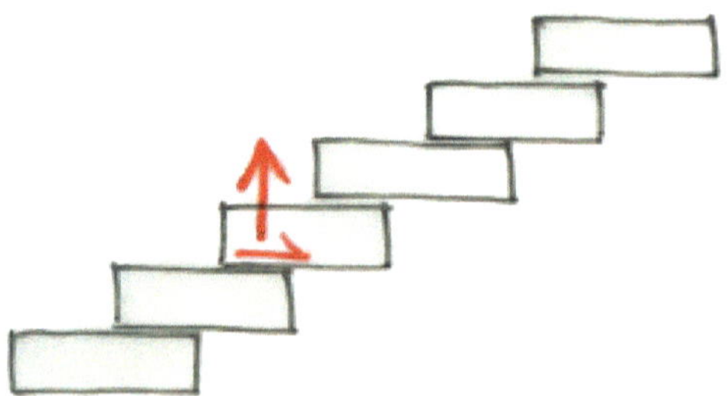

13.6.4 Using floors to bridge missing walls

Floor construction becomes relevant when using the notional removal approach in a masonry structure. As explored in Case study 11, a precast plank floor with a screed could have the capacity to span in the opposite direction to that of the plank, if a supporting wall were removed, with the screed doing the work.

A two-way spanning, *in situ* concrete floor may also be able to span in a single direction if a supporting wall is blown out. However, a timber floor will not be able to perform in this way. This is why floor construction and layout is so important when considering the disproportionate collapse strategy of an existing masonry building.

The structural engineer should not expect that a robustness strategy that might work for one masonry building will work for another. Key to working with existing masonry buildings is understanding the actual structural behaviour of each specific building.

13.7 Case studies — vertical extensions and other alterations

A significant consideration in vertical extension schemes is whether the existing structure is occupied or unoccupied, as access and design is simplified when a building is fully accessible. Case studies 17–23 include approaches to buildings which are both occupied and empty, and also include buildings where a change of use is proposed.

The studies are mostly a suggestion of the thought-process that an engineer would go through; it would be expected that the design engineer would also follow-up with detailed design and drawings and assessment of the structural performance.

Case study **17**

Consequence class change from CC2a to CC2b

Loadbearing wall structure (Figure 13.29)

The aim here is to consider the approach to take when a building changes between consequence classifications (CC2a to CC2b).

Figure 13.29: Structure-specific risks identified

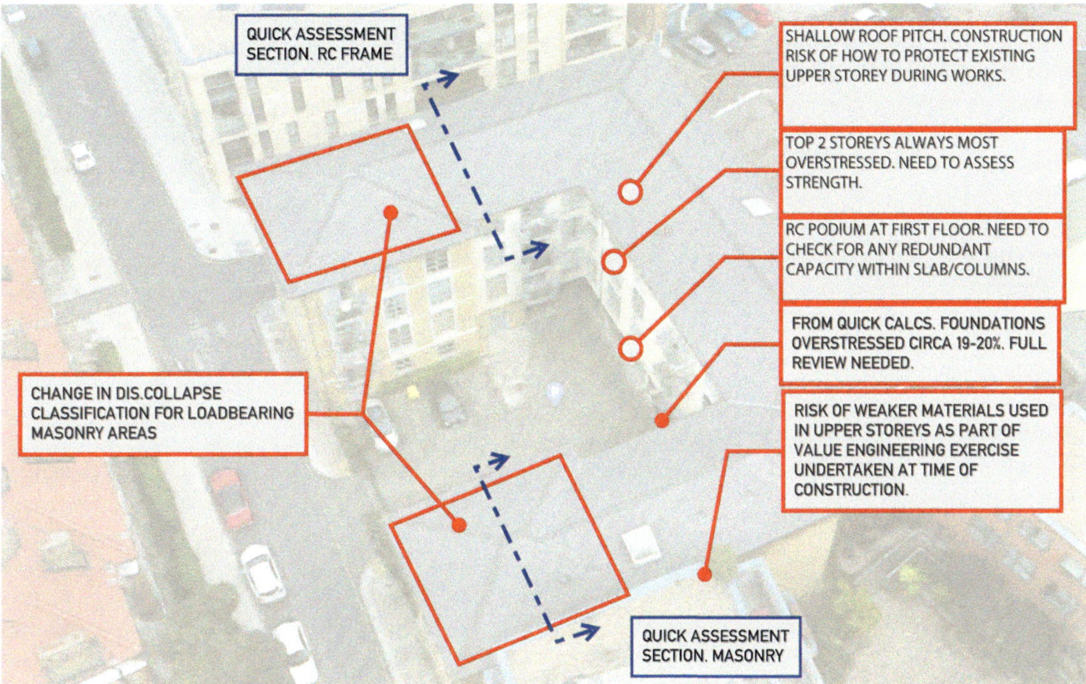

Generally the process should be as follows:

- Visual condition survey
- Investigation and testing
- Feasibility report, which includes but is not limited to:
  - Disproportionate collapse strategy
  - Review of the scheme and identification of key risks
  - Quick stress check (less than 10% increase)

The feasibility report should look to address:

- CDM regulation assessment regarding construction method
  - Difficulty in removing the roof, adding the transfer structure and maintaining a weathertight structure included within the key risk
- Providing evidence that the proposed scheme is achievable (i.e., stress and strength checks (Figure 13.30))
  - Existing calculation and drawing package reviewed to confirm loads
  - Independent load take-down made
  - Detailed calculation check complete with unity values also compared against quick calculations

Figure 13.30: Detailed check against initial quick stress check and existing calculations

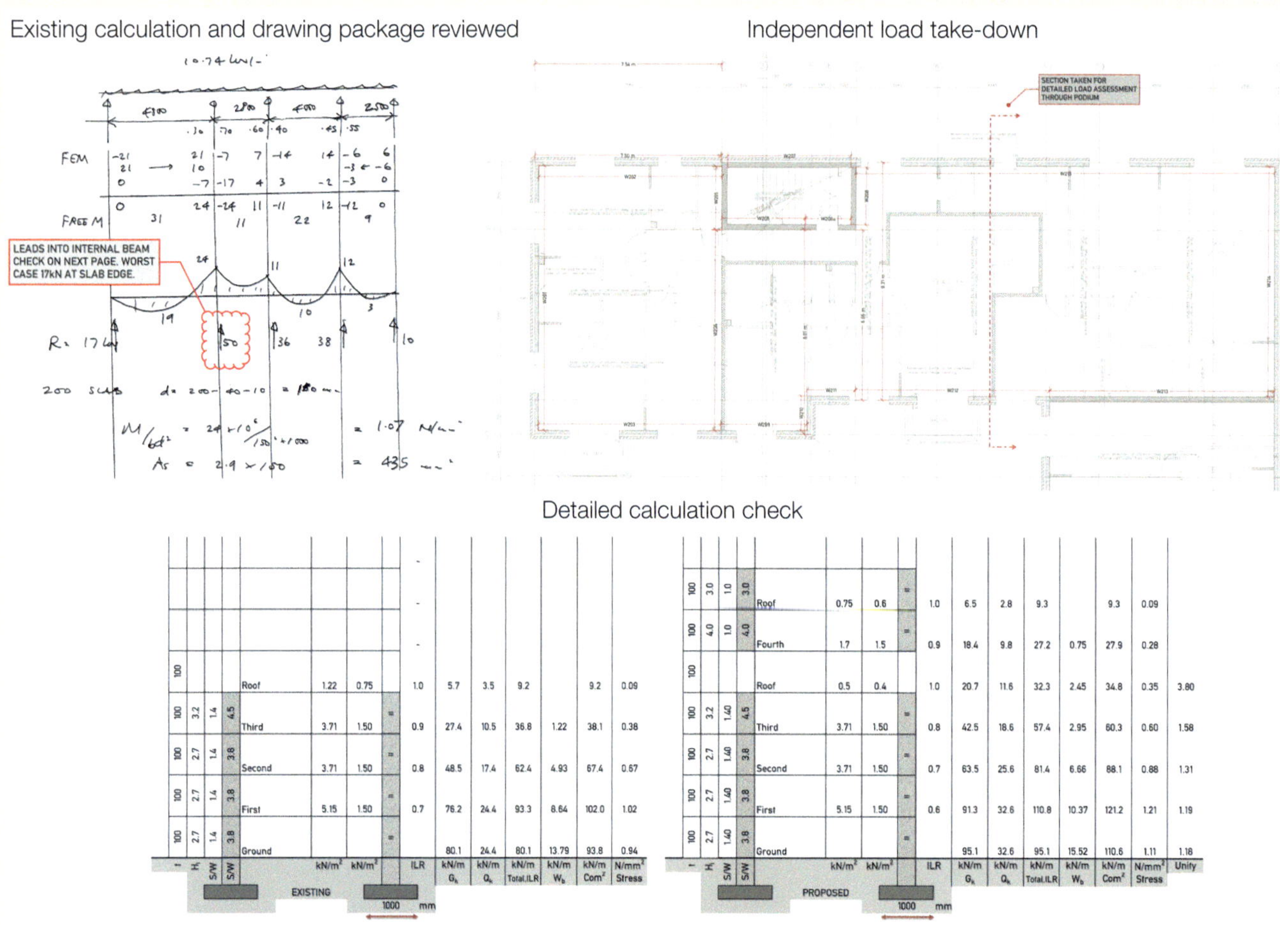

- Confirm disproportionate collapse strategy
  - Steel ring beam/grillage system adopted across the entire rooftop structure, which will act to span over the existing structure in the event of an accident in the storey below. New structure designed as CC2b
  - Notional element removal approach taken for masonry (Section 13.2.1) to check compliance with A3 as a CC2b building, and collapse areas checked to be within allowable limits

- Eliminate any risks to the existing structure showing extent of strengthening works to be carried out, where applicable (the design of which can be included within stage 4 but specialist strengthening work items should have been approached and confirmed as achievable as part of the feasibility report stage)
  - Strengthening works reviewed for +2 storey option (if pursued) and steel tonnage estimated. Beams to be strengthened are highlighted in red in Figure 13.31

- Provide sketch/plan of required transfer structure (grid/column layout) between proposed and existing structure (Figure 13.32). This should be taken to stage 3 (member sizing) which allows the developer to undertake a costing exercise of the required works prior to adding the new-build element of the design

Figure 13.31: Areas for strengthening identified

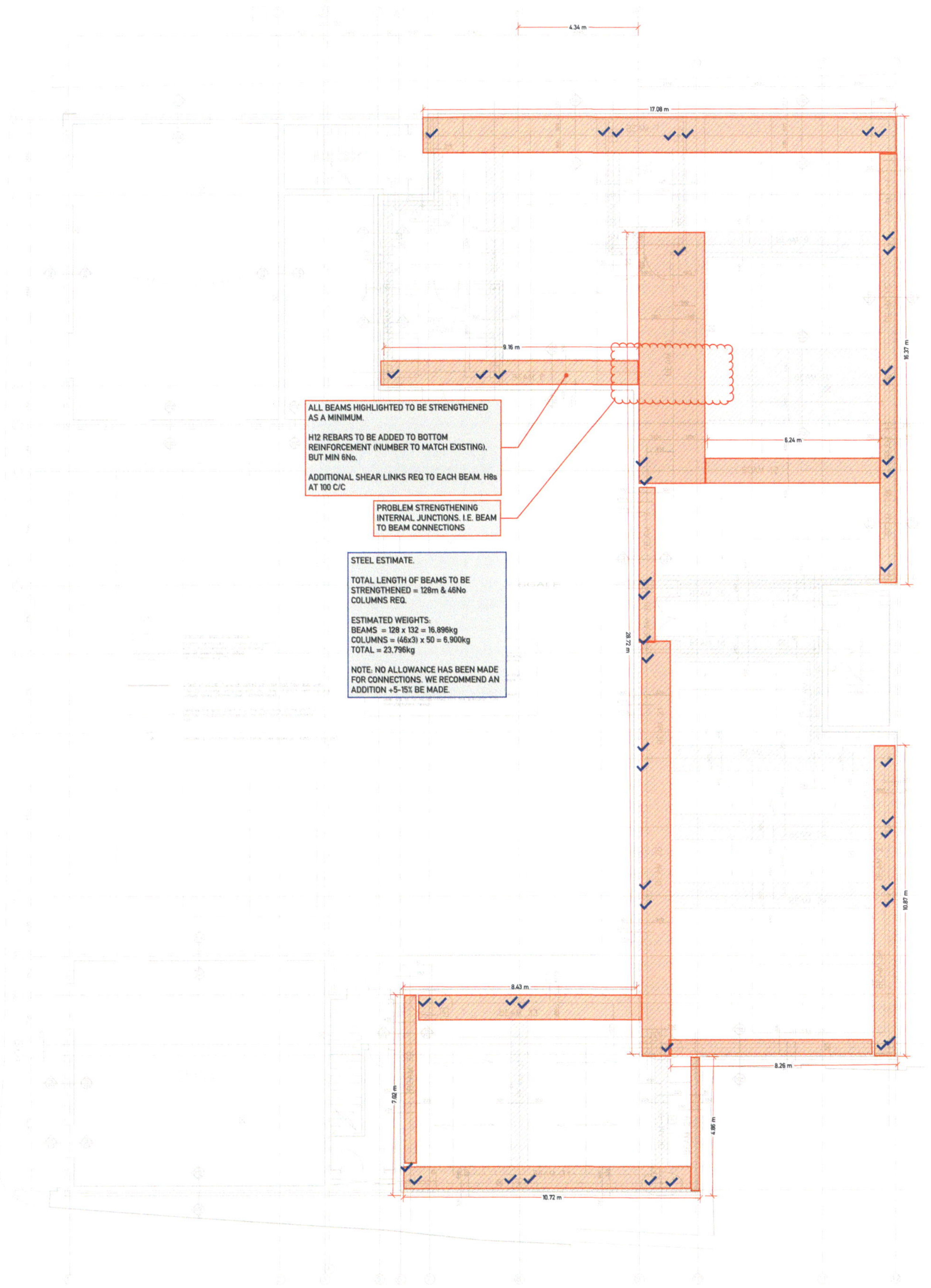

Figure 13.32: Indicative transfer deck arrangement

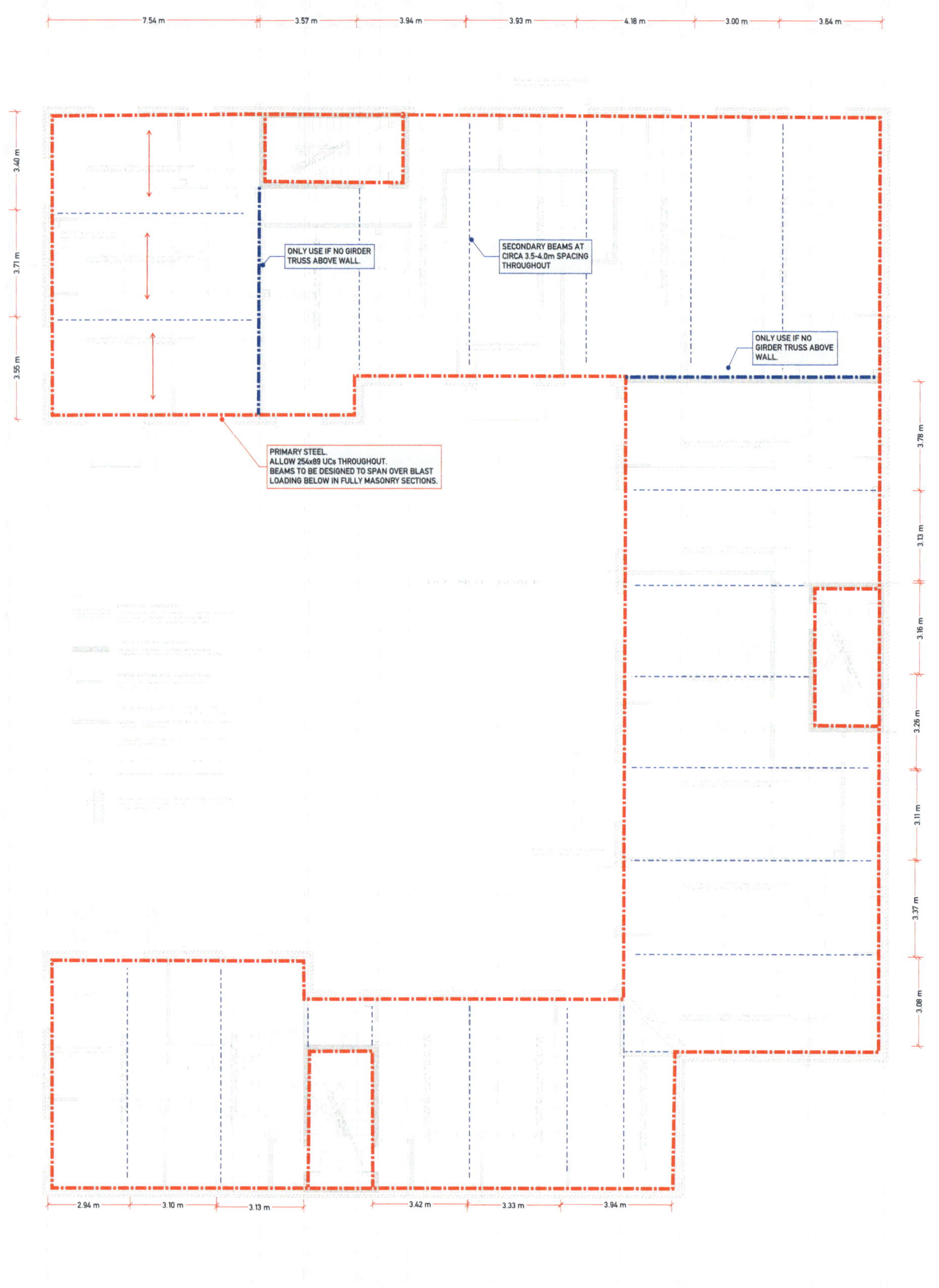

Case study 18

Framed structure with proposed additional storey

Five-storey concrete-framed building — single-storey steel-framed structure to be added on top
The aim is to understand when to delay a project. This is an example of a scheme proceeding through planning without any prior engineering input.

A five-storey reinforced concrete-framed residential apartment block built in 2010, which has planning approved for an additional storey. Proposed architectural layouts are included, but there has been no structural engineering input prior to the planning stage for the proposed vertical extension.

Recommended approach:

- Visual condition survey — N/A
- Investigation and testing — N/A
- Feasibility report, which includes but is not limited to:
   - Disproportionate collapse strategy
   - Review of the scheme and identification of key risks
   - Quick stress check

The building is occupied, therefore the engineer concentrates on the last three points first, looking at the following:

Disproportionate collapse strategy and options:

- Current disproportionate collapse consequence classification: CC2b
  The existing building falls under current building regulations CC2b. There is a concrete frame throughout and existing drawings/calculations show vertical and horizontal ties provided through walls/columns and slabs

- Proposed disproportionate collapse consequence classification: CC2b
  The 275mm-thick existing roof slab is proposed to be used as a new floor. Therefore the new frame will have to match the existing column/wall layout below.

- Proposed disproportionate collapse solution: Align the steel columns with the existing column/wall positions below. There are two possible options for compliance:
   - Where lift overruns or similar upstands occur, break out the existing upstand and adjust existing ties so that a new connection is formed with the wall/column. This continues the ties from foundation to roof (Figure 13.33)
   - Retain the roof, aligning the new columns with those below, and the new steel frame is then to be designed for CC2b, maintaining overall building classification. However, there is no vertical tie connection between the new steel columns and existing concrete columns. Therefore, provide an alternate load path in the event of an accident by designing the roof beams to be able to support the columns under, should they need to hang from the roof in the event of an accident below (Figure 13.34). The existing roof slab is checked under accidental loading ($34kN/m^2$) and debris loading

Figure 13.33: Structure-specific risks identified

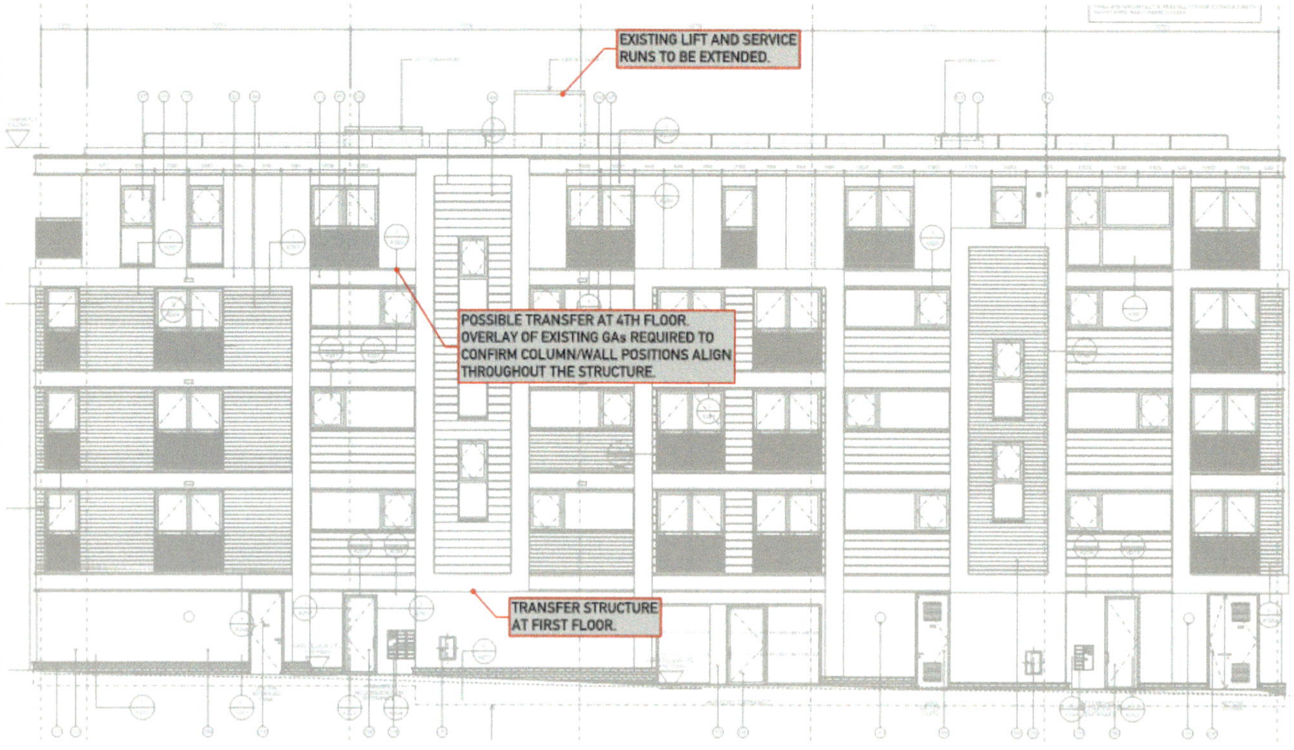

Figure 13.34: Proposed layout over existing structural grid below

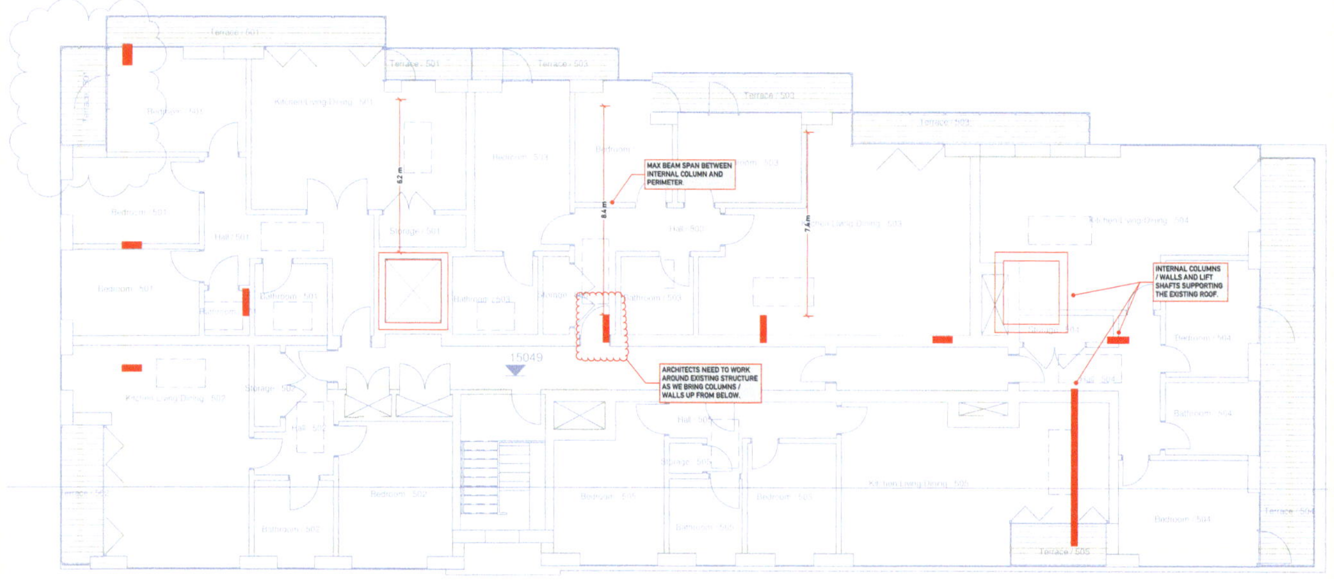

The new floor layout does not enable columns/walls to align:

- Alternative disproportionate collapse solution. Existing roof must act as a transfer structure:
  - Existing roof loads are fairly uniform area loads — lightly loaded
  - Proposed area loading increases (from roof to residential imposed loading) along with the prospect of line and point loads from wall/column structure supporting the new roof above

- Lack of current make-up information on the existing roof slab would require a new transfer deck to be provided to enable the proposed scheme. This would increase the height of the building by approx. 1m. This is a height (and cost) allowance that has not been made within the proposed elevations

- The existing photovoltaic panels on the roof will need to be moved up, increasing the weight of the proposed roof

- The spans between existing columns/walls are considered for roof beam sizing and intermediate joists/slabs. Secondary steelwork is assumed to be required to break long spans

- The existing structure will include anywhere between a 20% and 100% increase in lateral load at each floor (stability system to be reviewed), but access for investigation and testing is restricted throughout the building and there is no access to the original design calculations/drawings

This scheme was returned to the client to review the proposed layouts and the impact on overall building height. The vertical load transfer could not be resolved because of the desire to retain the existing roof slab, despite there being no information about its existing make-up.

Areas were inaccessible for investigation and testing so the existing structure could not be properly assessed, and the original calculations/drawings could not be sourced.

The scheme was not considered viable until further information could be found and made available to the structural engineer.

Case study 19

Investigation and testing

A four-storey mid-terrace property in central London
The aim is to show the benefit of pursuing further investigation and testing.

This building is a mixed-occupancy residential building with an apartment on each floor (Figure 13.35). The freehold owner wishes to add an extra storey to the building. The building is of traditional construction with timber floors, a central spine wall and a rear extension.

Figure 13.35: Typical elevation and section

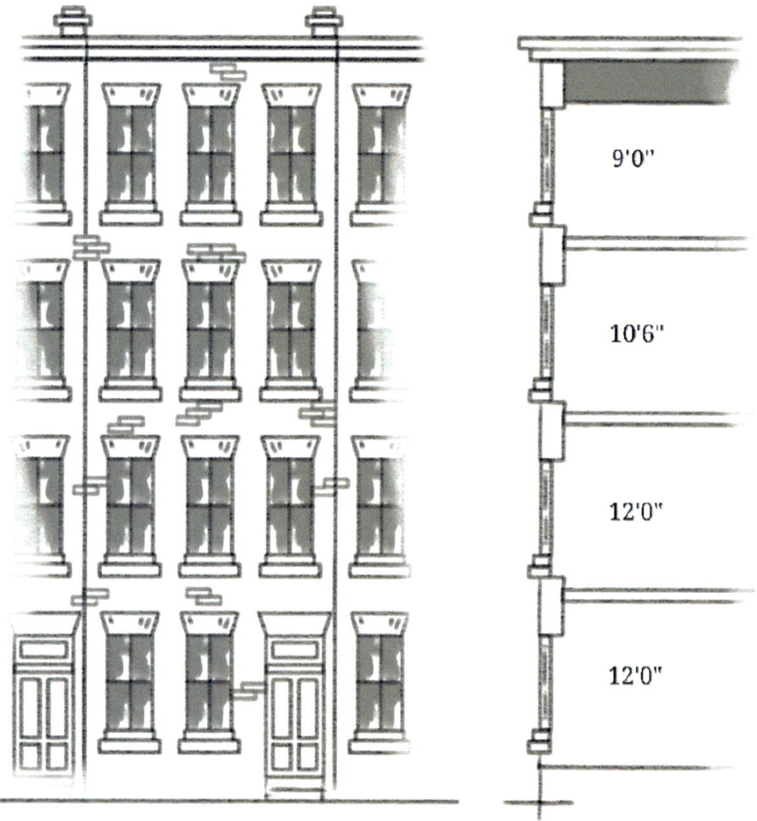

The additional storey prompts building work (Clause 3 (1) (a)) and material change of use (Clause 5(g)).

Recommended approach:

- Visual condition survey
- Investigation and testing
- Feasibility report, which includes but is not limited to:
    - Disproportionate collapse strategy
    - Review of the scheme and identification of key risks
    - Quick stress check

The building is occupied, therefore the structural engineer concentrates on the last three points first.

Disproportionate collapse strategy:

- Current disproportionate collapse classification: CC2a
  The existing building was constructed prior to current building regulations and is likely to be non-compliant with A3 of the Building Regulations

- Proposed disproportionate collapse classification: CC2b
  Clause 5, Building Regulations. Change in classification prompts building work and the structure to be made compliant. However, there is no access to the existing structure for strengthening works. It is necessary to ensure that the existing structure *"is no more unsatisfactory in relation to that requirement than before the work was carried out"*

  It is decided that a steel ring beam will be adopted along external elevations and secured to party walls, to provide the new structure with the ability to span over existing structure below in the event of an accident

Review of the scheme and identification of key risks:

- The building is old, with expected corbelled (shallow) foundations
- It is unlikely that the existing building even meets the requirements of CC2a. In addition, there may be deterioration of solid timber floors that are embedded into solid masonry walls
- Access is limited to the property. Although the whole property (freehold and leaseholds) is owned by one company, the property is tenanted
- The building has been altered in the past, with walls removed and (assumedly) replaced with steel beams, compromising the stability of the building

Quick stress check:

- The check suggests that the scheme is overstressed and investigation and testing would be required to pursue the scheme (Figure 13.36)

Figure 13.36: Quick stress check

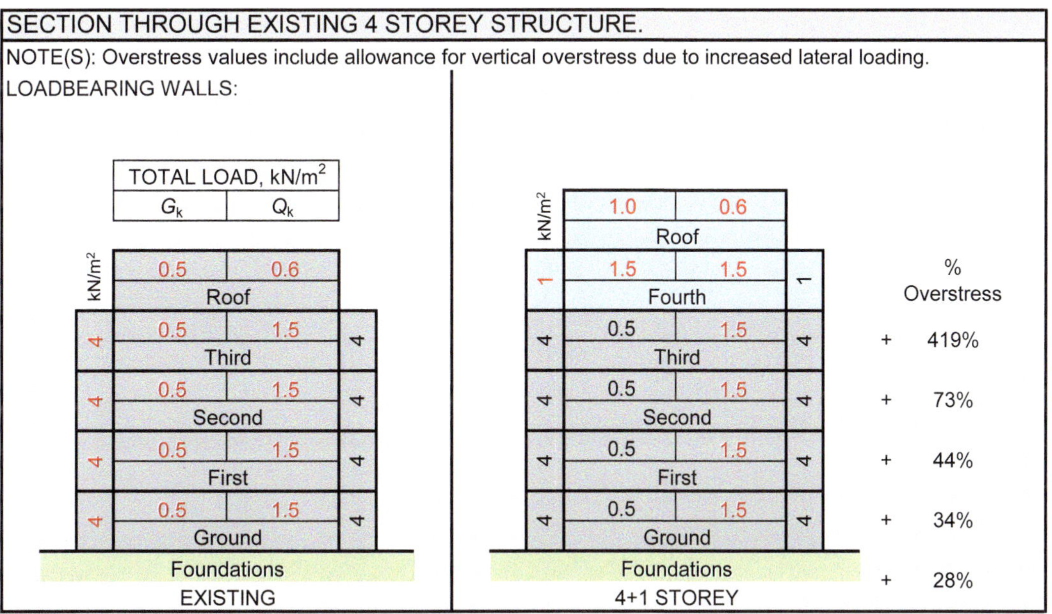

Comments:

The property is located in Kensington and the value of a single apartment would be incredibly high. The owner believes that the project will still be viable if all tenants are given notice to leave to allow for investigation and testing.

Investigation and testing:

Once the building is empty, the design engineer is able to do a full visual condition and investigative survey of the property. The engineer identifies the load paths and can confirm that very few structural alterations have been done to the property over previous years, with loadbearing walls continuous through each storey. The floor joists are found to be built into the walls, a detail that is not appropriate for CC2a or CC2b buildings.

The feasibility report should include:

- CDM regulation assessment regarding construction method
  - The property has been vacated, so risks specific to the proposed construction method should be highlighted on the general arrangement drawings

- Evidence that the proposed scheme is achievable (i.e., stress and strength checks)
  - The engineer is able to undertake trial holes of the foundations and confirm that the foundations are deep enough and the ground sound enough for loading to the structure to be increased
  - Material sampling of the existing masonry confirmed sufficient strength and capacity for an increase in loading

- Confirmation of the disproportionate collapse strategy
  - A building which does not satisfy the requirements for disproportionate collapse is at risk. It is also illegal because of the Statute requirement that a building *"is no more unsatisfactory in relation to that requirement than before the work was carried out"*
  - It is proposed that existing timber floors are retained but enclosed within a steel ring beam in each space. The steel ring beam acts as a perimeter tie (end plates between steel sections designed for tie force) resin-anchored into the masonry
  - Additional robustness can be gained by following CIRIA C579[132] recommendations for installing anchors into masonry at an angle instead of perpendicular to the wall. This is seen as good practice rather than quantifiable
  - The walls are checked to confirm that they can be removed in the event of an accident. The spine wall, front and rear walls are notionally removed and the consequence assessed. The steel ring beams can span over the missing wall element and all is found to be acceptable
  - The engineer proposes that the new storey is built on a grillage of steel beams with lightweight floor, wall and roof construction to keep the additional weight as low as possible

- Elimination of any risks to the existing structure, showing extent of strengthening works to be carried out where applicable. The design of any strengthening works can be included within stage 4 but specialist strengthening work items should have been approached and confirmed as achievable as part of the feasibility report
  - The engineer concludes that the building will be more robust (even with an extra storey) than the unaltered building. Tests of various scenarios of walls and floors being blown out have confirmed that the collapse would be minimal

- Sketch/plan of required transfer structure (grid/column layout) between proposed and existing structure. This should be taken to stage 3 (member sizing) which allows the developer to undertake a costing exercise of the required works prior to adding the new-build element of the design

Case study 20

Resisting pressure from a developer

A four-storey terrace in central London

The aim is to highlight the importance of resisting developer pressure.

The design engineer is asked to look at a similar building to that detailed in Case study 19. This building is located in a more modest London suburb. The building is empty, which gives the engineer an opportunity to do a visual and investigative survey. It is discovered that many of the internal spine walls have been demolished over the years, and that the timber floors and masonry are in poor condition.

The design engineer proposes a similar solution to that of the building in Case study 19, but suggests steel frames and new foundations to replace the walls that were removed, designed for the lateral loads on the building, making the assumption in the assessment of wind loading that the building is stand-alone.

The developer queries this, stating that wind loading on the side of the building is minimal due to the building being part of a terrace. The engineer is uncomfortable reducing wind loads because there is the strong likelihood that other buildings in the terrace have also had their stability compromised.

The developer accepts that the cost of the proposed works to the building, including the steel frames, steel ring beams to the floors and additional storey do not make the project cost-effective.

Engineers must be confident to say 'no' to developers where structural integrity is compromised.

Case study 21

Material change of use

A nine-storey steel and masonry structure

The aim is to review the building to ensure correct compliance for material change of use.

An existing building in Birmingham city centre, built *circa* 1930, is formed using solid masonry external walls with steel beams and columns framing the walls internally that support a hollow pot floor and roof system.

The structure is currently used as an office and commercial premises (Figure 13.37). Its classification is CC2b but it is non-compliant due to its age. A developer is looking to purchase the site and convert the existing structure into a boutique hotel.

Figure 13.37: Typical floor plan with area and measurements shown

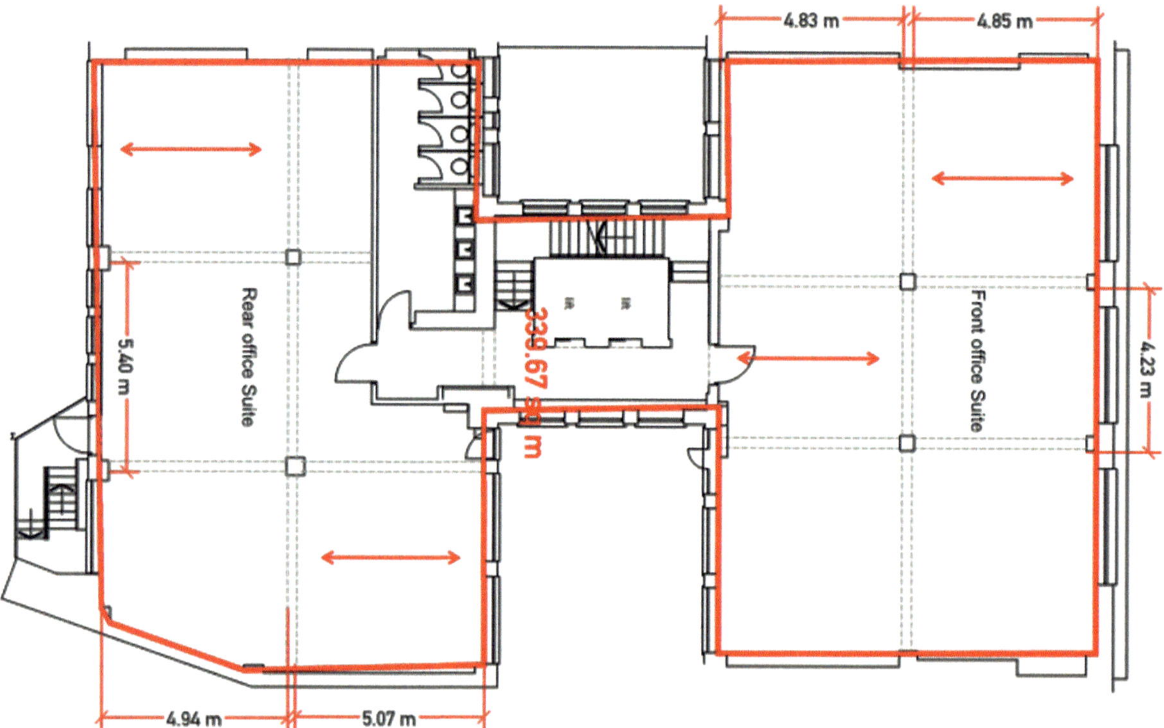

The new consequence class is also CC2b. However, under Clause 4 of the Building Regulations, it also falls under a material change of use, which then triggers building work and the requirement for the structure to comply with Schedule 1, which includes disproportionate collapse requirements.

Investigation and opening-up works show that the internal columns are steel-framed, brickwork encased. Incoming steel beams bear into steel cleats at the column and onto padstones within the external/perimeter internal walls. As the building will be empty during handover the client has the opportunity to make the building compliant, by providing vertical ties beneath each of the steel beams (from roof down to foundations).

Alternative solution:

Approved Document A also states that notional wall removal can be employed for CC2b buildings. 2.25h is taken as the notional opening length to be created in the external envelope. Therefore:

2.25h x (largest floor span) = area of floor blown out (A_{blow})

(A_{blow} ÷ 15) x 100 = required total area of the floor to fall within 15% Approved Document A limits

This area is compared to the actual floor area for compliance. This approach assumes that the masonry arches over the opening, and the floor below is checked for debris loading. In this example it is known that the proposed floor area that collapses is approx. 27.5m^2 which is less than 15% of the total floor area (approx. 340m^2).

In this approach, the contribution of the partitions has been ignored even though BS EN 1991-1-7 states:

"Clause A.7(1). The nominal length of loadbearing wall construction referred to in Clause A.4.(1)c should be taken as follows: For an external masonry, or timber or steel stud wall, the length measured between lateral supports provided by other vertical building components (e.g., columns or transverse partition walls)"

While the close-centre positioning of hotel bedrooms would have reduced the 2.25h length of wall taken into consideration, the practical reality is that the partitions will be non-structural, easily removed at a later date and are likely to provide little to no resistance in the event of a blast load. Adhering strictly to the codes here would have been counterproductive when considering structural safety and robustness. Therefore when considering alterations to existing masonry building stock consider:

"The spread of damage can also be limited when the intersecting wall is a substantial partition. This is a special instance of the previous case. In order to fulfil the requirements, the partition must:

- *Intersect the wall at right angles*
- *Average weight ≥ 150kg/m^2 — a half brick wall would be adequate here*
- *The junction must be capable of transmitting a tensile force, F_t, of 0.5F_t/m height or greater. Where the blast load (34kN/m^2) is applied to the loaded area and the resultant tensile force is resisted between wall and partition junction"*[130] (Figure 13.38)

Figure 13.38: Assessing the load on a partition junction

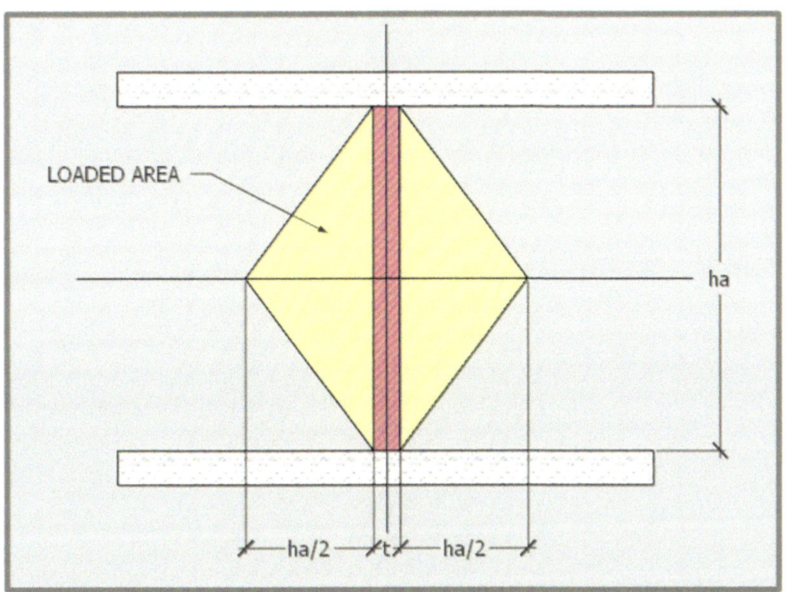

Case study 22

Material change of use and change in consequence class

Two-storey single-occupancy house converted into flats
The aim is to consider material change of use and also change in consequence class.

It is proposed that a two-storey single-family dwelling is converted into apartments (Figure 13.39). The original consequence class for the building is CC1, however, once it is converted to apartments it falls into CC2a.

The existing dwelling is an older property with floor joists that are built into masonry walls but without restraint straps or joist hangers. This is a detail that satisfies CC1 requirements but not CC2a.

Figure 13.39: House to be converted into apartments

As well as a change in consequence class, the proposed works also count as a material change of use. Clause 5(b) of the Building Regulations states:

"For the purposes of paragraph 8(1)(e) of Schedule 1 to the Act and for the purposes of these Regulations, there is a material change of use where there is a change in the purposes for which or the circumstances in which a building is used, so that after that change…

   (b) the building contains a flat, where previously it did not."

However, Clause 6 of the Building Regulations does not expect this particular change of use to comply with A1–3 of Schedule 1, but since the definition of building work includes 'material change of use', the engineer must consider the requirement:

"…complies with the applicable requirements of Schedule 1 or, where it did not comply with any such requirement, is 'no more unsatisfactory in relation to that requirement than before the work was carried out'."

Although not required, it is feasible to make an existing masonry and timber building compliant as a CC2a building, because effective anchorage of suspended floors can be achieved with a detail similar to that shown in Figure 13.40.

Figure 13.40: Effective anchorage detail

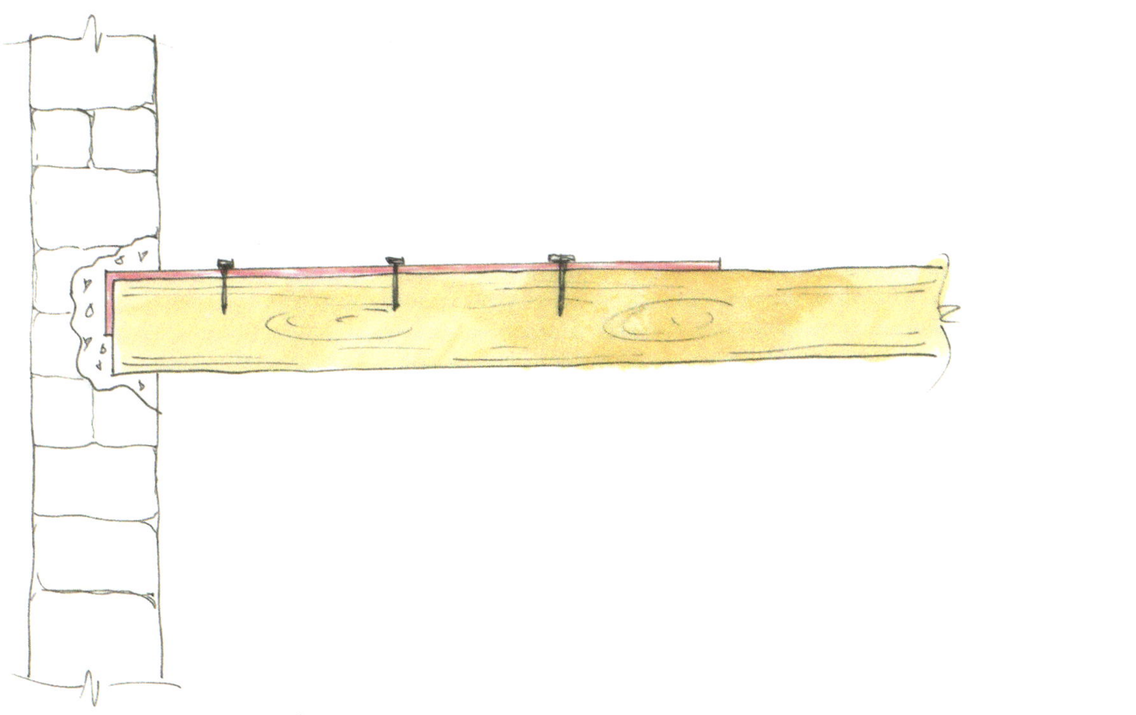

Case study 23

Change in consequence class and significant internal alterations

Five-storey dwelling converted into flats

The aim is to consider a change in consequence class with significant internal alterations due to multiple internal walls being removed.

The existing building is of masonry construction with timber floors. It is an old building, *circa* 1790, and the floor joists are built into the walls (Figure 13.41).

The original building class is CC2a and the proposed building classification is CC2b.

Figure 13.41: Five-storey dwelling

Strapping is not provided at any floor level and therefore the details do not comply with either CC2a or CC2b.

In addition to the change in building class, whole or parts of walls are being removed at all levels to enable layout changes. For this to be achieved with respect to vertical and horizontal load transfer, the design engineer is planning to include picture frame-type structures from foundation level up to the eaves.

The design engineer concludes that these new steel structures can act as horizontal and vertical ties, and also support floors should any lengths of wall be lost.

In light of the significant works to the building, the engineer carefully investigates the building, checking the impact if external and internal walls are removed between lateral supports at each floor level.

Cross Safety Report: Scottish tenements[133]
This report addresses concerns from a structural engineer about the removal of walls in Scottish tenement buildings and other alterations. The report is from 2006 but the issues are still relevant. The report points out the limitations of many goal post and picture frame designs.

CROSS Safety Report: Progressive collapse of an old mill building during a fire[134]
This report covers the sudden, swift, progressive collapse of an old mill building during a fire.

The Building Regulations 2010[26]

Party Wall etc. Act 1996[104]

The Construction (Design and Management) Regulations 2015[13]

Building Safety Act 2022[2]

Design guidance for strengthening concrete structure using fibre composite materials[126]

14 The risk of disproportionate collapse in fire

14.1 Introduction

The fire resistance design approach is commonly used throughout the world, but the history and limitations of this approach are not widely understood among structural engineers. This could lead to the disproportionate collapse of structures in fire.

There is a growing gap between the straightforward traditional structures that the elemental fire resistance design approach could be considered adequately to represent, and the actual design aspirations of today, such as long-span steel frames, modern methods of construction (MMC), mass timber and reuse of existing structures. It is therefore important that structural engineers understand what fire resistance is, what it is not, and what that means for demonstrating structural robustness in fire.

Structural engineers need to consider fire as a design load case, and understand if the simple, element-by-element fire resistance design approach (e.g., following fire safety design guidance in Approved Document B[101], BS 9999[135] or BS 9991[136] and the Eurocode approaches for member design/protection specification) is appropriate for a particular building and structural form.

For successful structural design for fire, close collaboration is needed between engineers, architects, building control, contractors, supply chain and others. Structural engineers and fire engineers must work together to determine where risk assessments and mitigations, more advanced structural analysis, potential design enhancements, and possibly the involvement of specialist structural fire engineering expertise may be necessary to evidence structural robustness to fire.

This chapter and the references and signposts to further reading, will support structural engineers in increasing their fire safety competency. This is essential to the industry's response to the Building Safety Act 2022[2], as well as evolving design aspirations in the pursuit of carbon reduction in the built environment which can lead to leaner structures.

14.2 Background

Structural engineers will be familiar with the term 'fire resistance'. It is probable that they will have worked with an architect to specify that steel members should be provided with passive fire protection (e.g., boarding, cementitious spray or intumescent paint) to achieve particular fire resistance periods. They may also have sized reinforced concrete members (and rebar cover) for these fire resistance periods, for example in line with Eurocodes (BS EN 1992-1-2[137]), or have sized some timber members for fire resistance (following BS EN 1995-1-2[138]). The required fire resistance periods would be set by the project fire strategy, following fire safety design guidance (e.g., Approved Document B, BS 9999 or BS 9991), or occasionally performance-based fire engineering routes.

The basis of this elemental fire resistance design approach and its relationship to structural performance in real fires is rarely well-understood by structural engineers or other members of the design and construction team. There are important assumptions and limitations both within this approach and also the standard furnace testing for fire resistance, and these are not widely-known or considered in the design stage. Applying these simplified concepts without due consideration of their limitations can undermine structural robustness in fire and/or result in outcomes that may not be aligned with expectations.

Significant structural collapses in fire are rare. In some major fires, including where fire has spread to multiple storeys, buildings have actually performed better or survived through fires of longer duration than the individual structural elements would have been tested or designed for. This is largely due to load redistribution. Examples include the fires in the First Interstate Bank, Los Angeles in 1988; One Meridian Plaza, Philadelphia in 1991; Broadgate Phase 8, London in 1990 and Torre Windsor, Madrid in 2005[139,140]. The Cardington fire test series conducted in the UK in the 1990s also provided a wealth of experimental evidence and insights into whole-frame composite steel–concrete structure behaviour in fire[141–143].

Fires of sufficient severity to threaten the structure are also comparatively rare events, and there are therefore limited opportunities to learn about structural robustness to fire. How much structural fire safety or structural robustness in fire is incorporated in typical buildings is impossible to be precise about. When buildings fail during or after fires, it is almost always because of structural behaviours that would not have been expected from standard fire resistance tests[144]. This is not surprising, as the furnace test was only ever intended as a comparative metric — it was never meant to represent real structural behaviour. It cannot be assumed that applying the elemental fire resistance design approach will guarantee adequate global structural performance in the event of fire in every building.

To consider fire as an accidental load case for robustness, as expected by BS EN 1991-1-7[4], structural engineers must be able to recognise when their designs may fall outside the standard elemental fire resistance design approach. This requirement is increasing, due to the widening gap between current design aspirations and underlying presumptions of that approach. For example, the imperative to reduce embodied carbon in construction is driving evolutions towards structural optimisation, lower carbon and new materials, lean construction and building reuse. Such shifts bring uncertainties and potentially introduce new and unknown failure modes, or erode conservatism within current or historic design approaches. There is therefore a need for proactive investigation and horizon scanning.

14.3 What is fire resistance?

The simplistic elemental fire resistance design approach is widely adopted across the world. It originated over 100 years ago, as part of combined efforts in the US, the UK and mainland Europe to create standardised testing regimes to compare the performance of loadbearing and fire-separating construction elements in fire, and the concepts have evolved relatively little since (Figures 14.1[145] and 14.2).

Figure 14.1: Exterior of test structure after a fire test

Figure 14.2: Example of a modern loaded fire resistance test for a structural element

For structural (loadbearing) fire resistance, elements are typically designed or protected to achieve particular fire resistance periods or ratings, expressed in hours or minutes. In fire safety design of buildings, structural stability and robustness are the final safety layer. Adequate structural performance must be maintained for a 'reasonable period' sufficient to safeguard escape, mitigate fire spread within the building and to neighbouring properties, and protect fire and rescue services personnel.

In most building codes, the required fire resistance periods for structural elements are assigned broadly-based on the building risk profile, which is correlated with factors such as building size, height and use. Elements designed or protected to meet higher fire resistance periods are generally assumed to perform better and more reliably in fire. This means that in higher consequence buildings (more people may come to harm in and around the building, longer evacuation times, protracted firefighting operations, etc.), higher fire resistance periods are generally recommended in design codes and associated guidance.

Some codes allow for lower fire resistance periods where automatic suppression (e.g., a sprinkler system) is provided, although it is important to note that this code relaxation is based on risk (due to the less-than-perfect reliability of sprinklers), because if the sprinkler system fails the fire resistance is the only feature remaining to protect the structure. Performance-based or risk-based approaches can also be adopted by fire engineers to determine an alternative appropriate fire resistance period for a building.

Traditionally, defining the appropriate loadbearing fire resistance period for a building in this way, has been considered adequate to satisfy the regulatory requirements. However, fire resistance is a test metric, and it is not intended to be a direct measure of real structural behaviour in fire. In addition, the prescribed periods of fire resistance given in hours or minutes in tabulated guidance (such as Table B4 of Approved Document B) rarely reflect the actual time to structural 'failure' in a real fire (or the duration of acceptable performance).

The history of fire resistance testing and its persistence as a core concept in building design and research even today are highlighted in Law and Bisby's 2020 paper[146], with further elaboration in their 2023 follow-up paper[147].

 *The rise and rise of fire resistance*[146]

 *Vive la résistance? Standard fire testing, regulation, and the performance of safety*[147]

Some of the key differences between fire resistance tests and possible behaviours in real building situations, with some relevant considerations, are:

- Test fire conditions are based on a 'standard fire', which is a suite of almost identical temperature–time curves used across the world[148–152]. The standard fire curve was originally devised in the early 20th century, informed by the experience of firefighters but also bounded by the limitations of furnaces. The continuously growing 'standard fire curve' may not be conservative in all cases. For example, the decay phase of a real fire would result in cooling of structural elements, which can induce significant additional tension forces in connections
- Test benchmarks of acceptable structural stability are based on arbitrary performance criteria, governed by constraints of the test arrangements (i.e., deflection limits to avoid damaging the fire test furnace) rather than any indicators of actual structural performance. Other acceptance criteria exist for fire-separating elements (Figure 14.3)

Figure 14.3: Primary fire resistance performance criteria for loadbearing elements (stability) and/or for fire-separating elements (integrity and insulation)

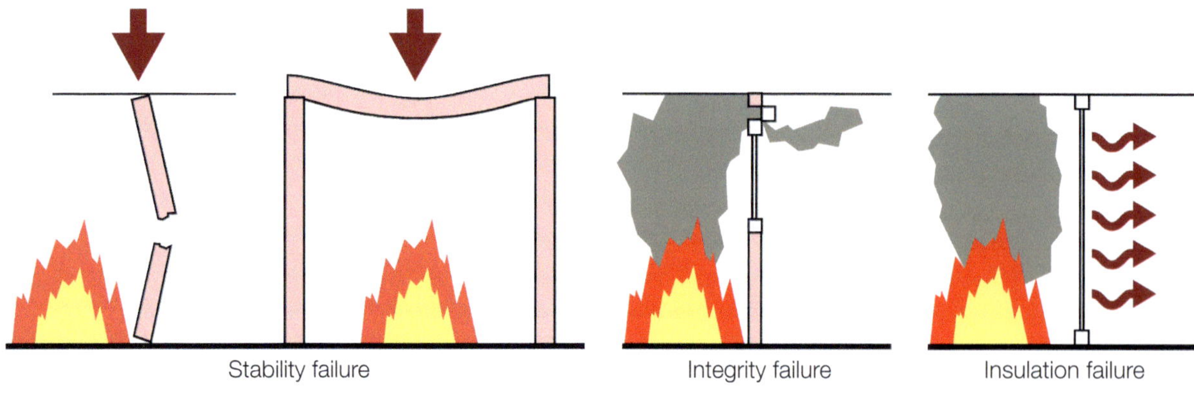

- Tests are typically on single elements (e.g., a beam, column or slab) with simplified end-support conditions and short spans (maximum of approx. 3m typically). These cannot represent the beneficial continuity, restraint, redistribution of loads, and possibly membrane action in real buildings. It also means they do not capture potentially detrimental second-order structural effects, such as eccentric loading and induced stresses due to expansion, contraction and displacement[153,154]
- The test fire conditions are not independent of the test specimen material. For example, more energy is required to sustain the gas temperatures when testing a concrete slab compared to a plasterboard-lined stud floor, on account of the differences in their thermal mass. In addition, the original tests were not intended for combustible enclosures, such as compartments with engineered timber panels (e.g., CLT). Any exposed timber surfaces on walls, ceilings or floors would contribute additional fuel load to the fire. This raises considerations about how the fire performance of elements tested in compartments with different linings (e.g., timber, concrete, masonry, plasterboard) can be credibly compared[99]

These limitations mean that many structural forms are not realistically represented by the standard fire resistance test. In addition, many of the tabulated design rules and simple calculation methodologies set out in standards such as the Eurocodes, have also been informed by the elemental fire resistance approach, and therefore also have limited ability to demonstrate the robustness of structural systems in real fires.

In the context of resilience and post-fire recovery (i.e., beyond only life safety requirements and regulatory compliance), the furnace testing and elemental fire resistance design approach do not give any indication of damage or repairability in the event of fire. For example, many structural members may need replacement after a major fire, even if they have not failed (e.g., steel beams that have undergone irrecoverable deformation).

There is a need for structural engineers to be more aware of the limitations of the elemental fire resistance design approach, and not to assume that it will always ensure robustness in ever-more-complex and optimised structures in the event of a fire.

14.4 Structural engineers and structural robustness in fire

Structural engineers need to engage with structural robustness in fire at all times. While specific requirements for structural fire performance exist in Part B of Schedule 1 of the Building Regulations (for England and Wales), fires are also accidents that have implications for compliance with Part A. Specifically, requirement A1(1) places an expectation on the structure to sustain loads and transmit them to the ground safely. Fires represent both an action on a structure, i.e., they have the potential to induce forces and moments, etc., and an impediment to the transmission of loads, i.e., through reductions in strength and stiffness with increasing temperature. Under requirement A3, fires can lead to collapses that are disproportionate to the cause. These implications of the regulatory requirements have not typically been well-understood or explicitly addressed by designers or even approving authorities.

Structural engineers have been encouraged to give greater consideration to performance in fire over many years — as can be seen in articles from *The Structural Engineer* archive spanning from 1968[155] to 2018[156], to more recent material around responding to the climate emergency[157]. This does not necessarily mean that detailed structural fire analysis is required for every project. Margaret Law MBE, the pioneering fire engineer, stated in her 1994 presentation and paper *Magic numbers and golden rules*[158], that:

"There should be a straightforward route for straightforward design"

For structural safety in fire, the 'straightforward' option could be the elemental fire resistance design approach. However, there is a need for competent professionals first to determine if this straightforward route is appropriate for a particular building, or whether alternative or more advanced approaches are needed. For this, the structural engineer needs to engage with the fire engineer, architect, building control, contractors, supply chain and others, to help all relevant parties to understand the structure, and together assess whether the elemental fire resistance design approach is appropriate for all parts of the structure.

The need for increased collaboration between fire engineers and structural engineers in particular, is made clear in the Building Safety Act, which requires a 'safety case' to be prepared for 'higher-risk buildings' (HRBs) where all relevant fire and structural safety hazards need to be considered.

To determinine whether the simple elemental fire resistance design approach is appropriate for a particular building and structure (or part of it), the following questions should be considered by the structural engineer, working with the fire engineer:

1. Does the furnace test heating regime adequately or conservatively represent the credible 'design fire' conditions for which the structure should be evaluated?
2. Will the furnace test or standard fire resistance design methods capture (or conservatively address) the possible structural failure modes in fire?

If the answer to either of these questions is 'no' or 'maybe', more consideration is required to demonstrate structural robustness in fire. This may involve further assessment or design work.

This assessment or design work will be required for a growing proportion of modern structures, as design evolution in recent decades has increasingly stretched beyond the assumptions of the standard elemental fire resistance approach.

Modern steel frames are notable cases, where long-spanning transfer beams, eccentrically loaded columns and inclined members in particular, present significant second-order effects due to large deflections and thermal expansion in fire[159].

It is also important to consider elements that may be particularly critical to structural robustness, and elements that are susceptible to non-ductile failure.

Some modern methods of construction (MMC) are also raising fire safety questions[160,161], including around structural robustness in fire, with some specific structural systems that may be more susceptible to disproportionate collapse in fire. A key example is light steel frame structures, where numerous, relatively small steel members run within the partition walls in the building, with the wall linings effectively forming fire-resisting 'shields' to the primary structure. As pointed out in CROSS reports[162,163], this presents single points of failure. Potentially large parts of the vertical loadbearing system could be compromised by relatively minor defects or damage to the passive fire protection, which in other structural systems may only affect one element locally.

Volumetric modular designs can have similar failure modes[164], although there may be some rigidity and load redistribution benefits to be gained from the block-like stacking of the modules.

Mass timber structure such as CLT slabs and walls or glue-laminated timber (glulam) beams and columns — which are increasingly being used in taller buildings — have been subject to increased attention in recent years[99]. These pose unique challenges for fire safety design, such as the impacts of fire dynamics and severity caused by exposed timber surfaces (on ceilings, walls and floors) and also that the structure itself is being consumed by the fire. The latter point is crucial in the context of robustness, as protracted and often hidden smouldering 'post-fire' and the resulting degradation of timber material properties due to continued heating through sections, can lead to collapse. The fire safety of timber structures is an evolving subject and those involved in the design of timber buildings should heighten their awareness of the challenges and current knowledge. Design guidance is provided in publications such as the *Fire safe use of wood in buildings — global design guide*[165]. Research and development on fire safety of timber structures has been increasing significantly in recent years, including large-scale fire testing[166–171].

14.5 Designing for structural robustness in fire

In its definition of 'robustness', BS EN 1991-1-7 states that fire should be considered as a load case, similar to other accidental loads like explosions, impact, or the consequences of human error. However, some of the usual ways to demonstrate robustness of structures against accidental loads may not be appropriate for fire.

The individual notional element removal approach would not reflect the fact that severe fires can affect many members simultaneously, unlike an explosion or vehicle impact, for example. Similarly, the key element approach — subjecting selected elements to an additional blast design load of $34kN/m^2$ — may not enhance fire resistance necessarily, and so would not demonstrate robustness in a fire. In addition, some of the typical detailing and design solutions employed for robustness cannot be assumed to deliver adequate resistance to disproportionate collapse in fire.

In a fire, the movement and restraint of the structure itself can be part of the 'load case', particularly in steel structures where deformations may be significant. Therefore, depending on the nature of the structure, a 'systematic risk assessment' such as recommended for CC3 buildings in Approved Document A is likely to be necessary for considerations of structural robustness in fire.

This should involve review of the relevant fire safety hazards and precautions, with the fire engineer (and other stakeholders as appropriate), with a view to reducing the likelihood of a major fire developing, for example following the ERIC hierarchy of risk reduction (Eliminate, Reduce, Isolate, Control). This process could also identify where enhancements to structural fire resistance or performance (i.e., keeping members cooler/stronger in a fire), or improving the ductility of the structure, may be beneficial to overall robustness, and where further structural calculations may be required.

It will also be important to explore where specialist structural fire engineering knowledge or analysis may be needed, to help evaluate or demonstrate robustness in fire. This could be at an elemental level, which may be appropriate if, for example, restraint conditions can be adequately captured. Examples of these typical approaches include the simple and more advanced calculation methods outlined in the Eurocodes (including National Annexes where applicable), notably for concrete, steel and steel–concrete composite elements[137,172,173]. Some similar methods are supported in other international guidance[174]. Alternatively, it could be at a structural system level, to help evidence structural robustness more explicitly (e.g., meet agreed performance criteria under particular credible 'design fires'). Further information on the field of structural fire engineering is provided in IStructE publications[175,176] and elsewhere[177–179].

Introduction to structural fire engineering[175]

Guide to the advanced fire safety engineering of structures[176]

Structural design for fire safety[177]

International handbook of structural fire engineering[178]

Fire safety engineering[179]

14.6 Fire robustness of existing structures in fire

The drive to prioritise retrofit continues to gather pace in the UK and globally, as part of efforts to decarbonise the built environment. Ensuring structural safety, is not only crucial to delivering effective retrofit projects, but also at the heart of early-stage feasibility and option studies, which can be the difference between choosing to demolish or reuse.

Reuse, adaptation and extension of existing structures pose particular challenges for structural robustness in fire. There are known shortcomings in both the robustness and fire resistance of some existing structures, when considered against modern regulations, codes and practices, which have evolved in response to some major structural failures, notably in relation to robustness design in the UK following the Ronan Point collapse in 1968.

There remain large panel system (LPS) buildings in the UK built in the 1960s before the Ronan Point collapse, which may have both poor robustness and also low structural fire resistance compared to current expectations, particularly of the slabs[6]. The potential for premature failure of a slab in fire presents a risk of disproportionate collapse, and a significant concern for life safety, particularly in residential buildings where there may be a 'stay put' fire strategy (i.e., UK fire safety guidance in Approved Document B and BS 9991) or prolonged evacuation processes.

To safeguard against disproportionate collapse in some older buildings, there may need to be increased reliance on preventing individual elements from failing. An example of this is recognised in the IStructE's *Appraisal of existing structures*[118], which recommends enhanced fire protection is provided to iron structures to keep them below 300°C (compared with modern steel structures where the limiting temperature might typically be in the 450–600°C range) unless further analysis of the global structure and connections is undertaken. This more onerous criterion is based on a research paper[180] which recommends keeping wrought iron beams relatively cool in fire, to minimise their deflection and expansion and to reduce the impact of eccentric forces being induced in brittle cast iron columns.

When working in existing buildings, it is important that structural engineers investigate the structure, understand its vulnerabilities (including to fire) at an elemental and global level, and then address these in remediation works, structural analyses and systematic risk assessments. There is a need for specific upskilling among structural engineers and fire engineers on the behaviour of existing buildings in fire.

The as-built structural fire resistance and structural robustness in fire needs to be understood by the fire engineer, so that it can be considered and integrated appropriately within the wider building fire strategy and precautions. Any residual risks or considerations must also be communicated to relevant parties so that they can be included in the fire risk assessment and ongoing operation of the building.

Case study 24

Delft University of Technology fire

On 13 May 2008, a fire in the 13-storey Faculty of Architecture Building at the Delft University of Technology in the Netherlands brought about a major structural collapse. It was a reinforced concrete building, made up of mainly precast inverted U-shape joists with *in situ* slabs between, spanning onto deep concrete girders and columns.

The fire began in a coffee vending machine on the sixth floor, quickly spreading to multiple floors in the building, which did not have any suppression system (e.g. sprinklers) installed. Although all building occupants evacuated safely, the rapid fire spread impacted firefighting operations, and the fire was allowed to burn uncontrolled for about seven hours, eventually resulting in structural collapse of a major part of the building — up to six bays, over the full height of the tower (Figure 14.4). The damage was ultimately significant enough for the building to be demolished.

Figure 14.4: Fire-damaged building showing area of collapse

The fire provided a rare opportunity to learn about major collapse due to fire in a high-rise reinforced concrete building. In the aftermath of fire, experts from Europe and the US led research on data collection and preliminary structural fire analyses, aiming to gain insights into what happened and what lessons could be drawn for structural fire design[181,182].

The preliminary hypothesis for the collapse was that large vertical deflections in the floor joists (due to loss of flexural capacity) resulted in catenary action and large axial tension forces being induced at the joist-to-girder connections, bringing about tensile failure of these connections, and ultimately precipitating buckling failure of the columns (Figures 14.5 and 14.6).

Figure 14.5: Typical floor plan for tower

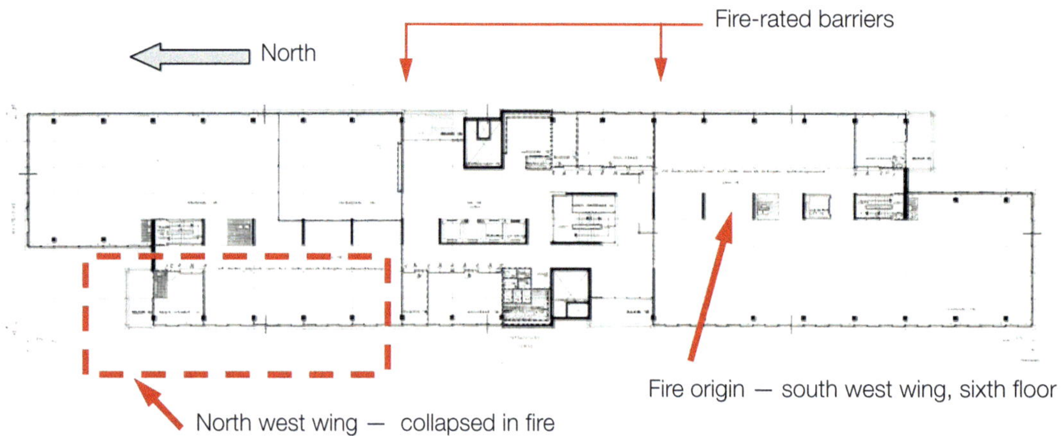

Figure 14.6: Major components of structural system

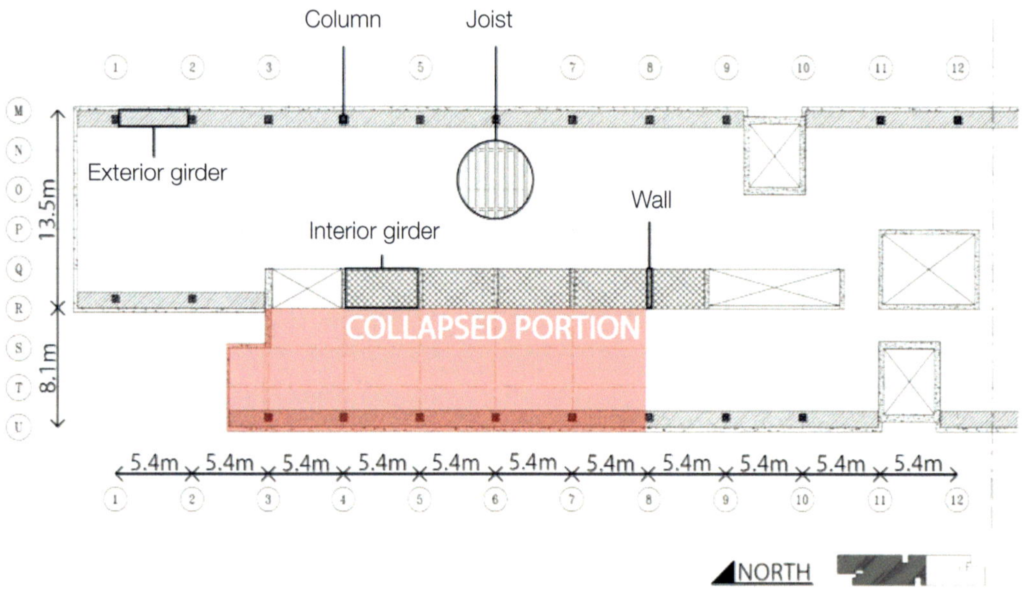

This structural collapse was a prime example of some of the key considerations outlined in this chapter. The key points relevant to the risk of disproportionate collapse in fire are:

- While collapse of major reinforced concrete structures in fire is infrequent, it is possible and warrants specific consideration by designers. There is a requirement, therefore, to take appropriate care in design and construction
- The loss of strength of individual members in isolation was not sufficient to explain fully the large-scale structural collapse. Although no full and detailed analysis of the structure as a system was undertaken, this fire event and the elemental analysis performed suggest the importance of taking a system-level approach to structural fire safety, which explicitly considers interactions of structural elements
- The fire moved through the building for about seven hours, with more than one floor typically burning at one time. When developing 'design fires' for structural fire engineering analysis of buildings, consideration should be given to the potential movement of large fires (particularly if, for example, there are large, open-plan spaces, if there are questions over the effectiveness of fire compartmentation in the building, and/or if there is no suppression system present) and how these may affect structural response
- The reinforced concrete members reached their lowest strength during the decay phase of the fire. Also, the collapse occurred as the fire was decaying out in the region of the collapse. This confirms that structures may be most vulnerable to collapse during the decay and cooling phase of a fire, rather than while the fire is burning most severely. This is a particularly important consideration for fire and rescue services, and for fire investigators

References

1   Hackitt, J. *Building a safer future: Independent review of building regulations and fire safety. Interim report.* Available at: https://assets.publishing.service.gov.uk/government/uploads/system/uploads/attachment_data/file/668831/Independent_Review_of_Building_Regulations_and_Fire_Safety_web_accessible.pdf. *Final report* available at: https://assets.publishing.service.gov.uk/government/uploads/system/uploads/attachment_data/file/707785/Building_a_Safer_Future_-_web.pdf [Accessed: November 2023].

2   HM Government. *Building Safety Act 2022.* Available at: www.legislation.gov.uk/ukpga/2022/30 [Accessed: November 2023].

3   The Institution of Structural Engineers. *Practical guide to structural robustness and disproportionate collapse in buildings.* London: IStructE Ltd, 2010.

4   British Standards Institution. *BS EN 1991-1-7:2006+A1:2014: Eurocode 1. Actions on structures — General actions — Accidental actions.* London: BSI, 2014.

5   CROSS UK. *CROSS Safety Alert: Structural stability/integrity of steel frame buildings.* Available at: www.cross-safety.org/uk/safety-information/cross-safety-alert/structural-stabilityintegrity-steel-frame-buildings [Accessed: November 2023].

6   CROSS UK. *CROSS Safety Report: Disproportionate collapse assessment of large panel system buildings.* Available at: www.cross-safety.org/uk/safety-information/cross-safety-report/disproportionate-collapse-assessment-large-panel-885 [Accessed: November 2023].

7   CROSS UK. *CROSS Safety Report: Risks from off-site manufacture and hybrid construction.* Available at: www.cross-safety.org/uk/safety-information/cross-safety-report/risks-site-manufacture-and-hybrid-construction-529 [Accessed: November 2023].

8   CROSS UK. *CROSS Safety Alert: Safety issues associated with balconies.* Available at: www.cross-safety.org/uk/safety-information/cross-safety-alert/safety-issues-associated-balconies [Accessed: November 2023].

9   CROSS UK. *CROSS Safety Report: Concern about design principles for tall buildings.* Available at: www.cross-safety.org/uk/safety-information/cross-safety-report/concern-about-design-principles-tall-buildings-235 [Accessed: November 2023].

10   CROSS UK. *CROSS Safety Report: Concerns over robustness of some 20 year old buildings.* Available at: www.cross-safety.org/uk/safety-information/cross-safety-report/concerns-over-robustness-some-20-year-old-buildings-1051 [Accessed: November 2023].

11   British Standards Institution. *BS EN 1991-1-6:2005: Eurocode 1. Actions on structures — General actions. Actions during execution.* London: BSI, 2005.

12   Temporary Works Forum. Available at: www.twforum.org.uk [Accessed: November 2023].

13   HM Government. *The Construction (Design and Management) Regulations 2015.* Available at: www.legislation.gov.uk/uksi/2015/51 [Accessed: November 2023].

14   CROSS UK. *CROSS Safety Report: Stability of tenants' mezzanine.* Available at: www.cross-safety.org/uk/safety-information/cross-safety-report/stability-tenants-mezzanine-378 [Accessed: November 2023].

15   HM Government. *Health and Safety at Work etc. Act 1974.* Available at: www.legislation.gov.uk/ukpga/1974/37 [Accessed: November 2023].

16   The Institution of Structural Engineers. *Temporary works toolkit.* 19 parts. Available at: www.istructe.org/thestructuralengineer/article-series/temporary-works-toolkit [Accessed: November 2023].

17   CROSS UK. *CROSS Topic Paper: FC Twente stadium roof collapse.* Available at: www.cross-safety.org/uk/safety-information/safety-information-cross-sub-heading-only/fc-twente-stadium-roof-collapse [Accessed: November 2023].

18   Heyman, J. 'Hambly's paradox: why design calculations do not reflect real behaviour'. *ICE Proceedings, Civil Engineering*, 114(4), November 1996, pp161–166.

19   The Institution of Structural Engineers. *Manual for the systematic risk assessment of high-risk structures against disproportionate collapse.* London: IStructE Ltd, 2013.

20   CONSTRUCT. *National structural concrete specification for building construction: fourth edition complying with BS EN 13670: 2009.* Camberley: The Concrete Centre, 2010.

21   British Constructional Steelwork Association. *National structural steelwork specification for building construction (BCSA Publication 62/20).* 7th ed. London: BCSA, 2020.

22   Building Research Establishment. *Report on the collapse of the roof of the assembly hall of the Camden School for Girls.* London: HMSO, 1973.

23   CROSS UK. *CROSS Safety Alert: Tension systems and post-drilled fixings.* Available at: www.cross-safety.org/uk/safety-information/cross-safety-alert/tension-systems-and-post-drilled-fixings [Accessed: November 2023].

24   Ministry of Housing and Local Government. *Report of the inquiry into the collapse of flats at Ronan Point, Canning Town.* London: HMSO, 1968.

25   HM Government. *The Building Regulations 2010. Approved Document A: Structure.* Available at: www.gov.uk/government/publications/structure-approved-document-a [Accessed: November 2023].

26   HM Government. *The Building Regulations 2010.* Available at: www.legislation.gov.uk/uksi/2010/2214 [Accessed: November 2023].

27   CROSS UK. *CROSS Safety Report: Collapse of Large Panel System (LPS) buildings during demolition.* Available at: www.cross-safety.org/uk/safety-information/cross-safety-report/collapse-large-panel-system-lps-buildings-during-186 [Accessed: November 2023].

28   Calvi, G.M. *et al.* 'Once upon a time in Italy: the tale of the Morandi Bridge', *Structural Engineering International*, 29(2), 2019, pp198–217.

29   CROSS UK. *Cross Safety Alert: The management of design related risks: structural civil and fire engineers.* Available at: www.cross-safety.org/sites/default/files/2021-12/cross_safety_alert_the_management_of_design_related_risks.pdf [Accessed: November 2023].

30   Department for Transport, Local Government and the Regions. *DTLR Framework Report: Proposed revised guidance on meeting compliance with the requirements of Building Regulation A3: Revision of Allot and Lomax proposal (Project Report Number 205966).* Available at: https://assets.publishing.service.gov.uk/government/uploads/system/uploads/attachment_data/file/429221/DTLR_Framework_Report.pdf [Accessed: November 2023].

31   CROSS UK. *CROSS Safety Alert: Hazard identification for structural design.* Available at: www.cross-safety.org/uk/safety-information/cross-safety-alert/hazard-identification-structural-design [Accessed: November 2023].

32   The Institution of Structural Engineers. *Risk in structural engineering.* London: IStructE Ltd, 2013.

33   British Standards Institution. *NA+A1:2014 to BS EN 1991-1-7:2006+A1:2014: National Annex to Eurocode 1. Actions on structures — Accidental actions.* London: BSI, 2014.

34   British Standards Institution. *BS EN 1990:2002+A1:2005: Eurocode. Basis of structural design.* London: BSI, 2005.

35   British Standards Institution. *PD 6688-1-7:2009+A1:2014: Recommendations for the design of structures to BS EN 1991-1-7.* London: BSI, 2014.

36   British Standards Institution. *BS EN 1993-1-1:2005+A1:2014: Eurocode 3. Design of steel structures — General rules and rules for buildings.* London: BSI, 2014.

37   British Standards Institution. *BS EN 1992-1-1:2004+A1:2014: Eurocode 2: Design of concrete structures — General rules and rules for buildings.* London: BSI, 2014.

38   British Standards Institution. *NA+A2:2014 to BS EN 1992-1-1:2004+A1:2014: UK National Annex to Eurocode 2. Design of concrete structures — General rules and rules for buildings.* London: BSI, 2014.

39   British Standards Institution. *BS 8110-1:1997: Structural use of concrete — Code of practice for design and construction.* London: BSI, 1997.

40   British Standards Institution. *BS EN 1996-1-1:2005+A1:2012: Eurocode 6. Design of masonry structures — General rules for reinforced and unreinforced masonry structures.* London: BSI, 2012.

41   British Standards Institution. *PD 6697:2019: Recommendations for the design of masonry structures to BS EN 1996-1-1 and BS EN 1996-2.* London: BSI, 2019.

42   British Standards Institution. *BS EN 1995-1-1:2004+A2:2014: Eurocode 5: Design of timber structures — General. Common rules and rules for buildings.* London: BSI, 2014.

43   British Standards Institution. *PD 6693-1:2019: Recommendations for the design of timber structures to Eurocode 5: Design of timber structures — General. Common rules and rules for building.* London: BSI, 2019.

44   British Standards Institution. *NA to BS EN 1990:2002+A1:2005: UK National Annex for Eurocode. Basis of structural design.* London: BSI, 2005.

45   Department for Communities and Local Government and Centre for the Protection of National Infrastructure. *Review of international research on structural robustness and disproportionate collapse.* Available at: https://assets.publishing.service.gov.uk/government/uploads/system/uploads/attachment_data/file/6328/2001594.pdf [Accessed: November 2023].

46   CROSS UK. Available at: www.cross-safety.org/uk [Accessed: November 2023).

47   HM Government. *Building Act 1984.* Available at: www.legislation.gov.uk/ukpga/1984/55 [Accessed: November 2023].

48   CROSS UK. *CROSS Safety Report: Concerns over risky new buildings?* Available at: www.cross-safety.org/uk/safety-information/cross-safety-report/concerns-over-risky-new-buildings-632 [Accessed: November 2023].

49   Russell, J.M *et al*. 'Historical review of prescriptive design rules for robustness after the collapse of Ronan Point'. *Structures*, 20, August 2019, pp365–373. DOI: https://doi.org/10.1016/j.istruc.2019.04.011 [Accessed: November 2023].

50   CEN/TC 250. *Mandate M/515. Development of 2nd generation of EN Eurocodes. Report of Project Team WG6. T2. Robustness rules in material related Eurocode parts. Final draft for informal enquiry. 2020-10-30*. [Brussels]: CEN, 2020.

51   Alexander, S. 'New approach to disproportionate collapse'. *The Structural Engineer*, 82(23), 7 December 2004, pp14–18.

52   NHBC. *Technical guidance note: The Building Regulations 2004 edition — England and Wales: requirement A3 — disproportionate collapse*. Amersham: NHBC, 2004.

53   Way, A.G.J. *Structural robustness of steel framed buildings: in accordance with Eurocodes and UK National Annexes (SCI Publication 391)*. Ascot, SCI, 2011. Available at: www.steelconstruction.info/images/0/0b/SCI_P391.pdf [Accessed: November 2023].

54   Standing Committee on Structural Safety (SCOSS). *The fire at the Torre Windsor office building, Madrid 2005*. Available at: www.cross-safety.org/sites/default/files/2005-06/torre-windsor-building-fire-madrid.pdf [Accessed: November 2023].

55   The Institution of Structural Engineers. *Temporary demountable structures: guidance on procurement, design and use*. 4th ed. London: IStructE Ltd, 2017.

56   Godart, B. and Datry, J-B. '2E terminal at Roissy-Charles de Gaulle Airport'. In Palmisano, F. and Rus. L. eds. *Case studies on failure investigations in structural and geotechnical engineering (IABSE Bulletins — Case Studies 4)*. Geneva: IABSE, 2023, pp61–85.

57   British Standards Institution. *NA+A1:2014 to BS EN 1993-1-1:2005+A1:14: UK National Annex to Eurocode 3. Design of steel structures — General rules and rules for buildings*. London: BSI, 2014.

58   Steel Construction Institute. *Vertical tying of columns and column splices (AD-415)*. Available at: www.steelconstruction.info/images/4/45/AD-415.pdf [Accessed: November 2023].

59   British Standards Institution. *BS EN 1993-1-8:2005: Eurocode 3. Design of steel structures — Design of joints*. London: BSI, 2005.

60   Steel Construction Institute. *NCCI: Tying resistance of a simple end plate connection (SN015a-EN-EU)*. Ascot: SCI, 2011.

61   Steel Construction Institute and British Constructional Steelwork Association. *Joints in steel construction: simple joints to Eurocode 3 (SCI Publication P358)*. Ascot: SCI, 2011.

62   Brettle, M.E. *Steel building design: worked examples — open sections, in accordance with Eurocodes and the UK National Annexes (SCI Publication P364)*. Ascot: SCI, 2009. Available at: www.steelconstruction.info/images/5/50/Sci_p364.pdf [Accessed: November 2023].

63   British Standards Institution. *BS 5950-1:2000: Structural use of steelwork in building — Code of practice for design. Rolled and welded sections*. London: BSI, 2001.

64   The Institution of Structural Engineers. *Car park design*. London: IStructE Ltd, 2023.

65   British Standards Institution. *PD 6687-1:2020: Background paper to the National Annexes to BS EN 1992-1, BS EN 1992-3 and BS EN 1992-4*. London: BSI, 2020.

66   The Institution of Structural Engineers. *Manual for the design of steelwork building structures to Eurocode 3*. London: IStructE Ltd, 2010.

67   British Standards Institution. *BS EN 1993-1-10:2005: Eurocode 3. Design of steel structures — Material toughness and through-thickness properties*. London: BSI, 2005.

68   British Standards Institution. *BS EN 1994-1-1:2004: Eurocode 4. Design of composite steel and concrete structures — General rules and rules for buildings*. London: BSI, 2005.

69   Yandzio, E., Lawson, R.M. and Way, A.G.J. *Light steel framing in residential construction (SCI Publication P402)*. Ascot: SCI, 2015. Available at: www.steelconstruction.info/images/2/23/SCI_P402.pdf [Accessed: November 2023].

70   Way, A.G.J. *Robustness of light steel construction (SCI Technical Information Sheet ED021)*. Ascot: SCI, 2014.

71   British Standards Institution. *BS 5950-5:1998: Structural use of steelwork in building — Code of practice for design of cold formed thin gauge sections*. London: BSI, 1998.

72   Lawson, R.M. 'Robustness of light steel frames and modular construction'. *ICE Proceedings, Structures and Buildings*, 161(1), February 2008, pp3–16.

73   Lawson, R.M. *Building design using modules (SCI Publication P348)*. Ascot: SCI, 2007. Available at: www.steelconstruction.info/images/6/67/SCI_P348.pdf [Accessed: November 2023].

74   The Institution of Structural Engineers. *Manual for the design of concrete building structures to Eurocode 2*. London: IStructE Ltd. 2006.

75   Federal Emergency Management Agency. *The Oklahoma City bombing: improving building performance through multi-hazard mitigation (FEMA 277)*. [s.l]: ASCE, 1996.

76   The Concrete Society. *Post-tensioned concrete floors: design handbook (Technical Report 43)*. 2nd ed. Camberley: The Concrete Society, 2005.

77   *fib. Design of precast concrete structures against accidental actions (fib Bulletin 63*. Lausanne: *fib*, 2012.

78   Billington, C. 'Achieving robustness of precast concrete stairs using proprietary cast-in inserts'. *The Structural Engineer*, 92(2), February 2014, pp34–36.

79   Brooker, O. *How to design concrete buildings to satisfy disproportionate collapse requirements (TCC/03/45)*. Camberley: The Concrete Centre, 2008.

80   Kelly, P. 'Robustness and disproportionate collapse'. *Spa News.*

81   CERAM Research. *The robustness of the domestic house. 4 Parts (Technical Notes 350, 364, 370, 371)*. Stoke-on-Trent: CERAM Research, 1983–1985.

82   Brick Development Association, Aircrete Products Association and Concrete Block Association. *Masonry design for disproportionate collapse requirements under Requirement A3 of The Building Regulations (England & Wales)*. Available at: www.brick.org.uk/uploads/downloads/masonry-design-for-regulation-a3.pdf [Accessed: November 2023].

83   British Standards Institution. *BS 5628-1:2005: Code of practice for the use of masonry — Structural use of unreinforced masonry*. London: BSI, 2005.

84   Curtin, W.G. *et al. Structural masonry designers' manual*. 3rd ed. Oxford: Blackwell, 2006.

85   Way, A.G.J. and Lawson, R.M. *Uninterrupted height of masonry cladding to light steel framing (SCI Technical Information Sheet P426)*. Ascot, SCI: 2019.

86   The Institution of Structural Engineers. *An overview of the specifying and detailing of masonry construction*. London: IStructE, 2019. Available at: www.istructe.org/resources/report/an-overview-of-the-specifying-and-detailing-of-mas/ [Accessed: November 2023].

87   UK Timber Frame Association. 'Timber Engineering Notebook series. No. 3: Timber frame structures — platform frame construction (part 1)'. *The Structural Engineer*, 91(5), May 2013, pp26–32.

88   British Standards Institution. *NA to BS EN 1995-1-1:2004+A2:2014: UK National Annex to Eurocode 5: Design of timber structures — General. Common rules and rules for buildings*. London: BSI, 2019.

89   Milner, M.W. *et al.* 'Verification of the robustness of a six-storey timber frame building'. *The Structural Engineer*, 76(16), 18 August 1998, pp307–312.

90   Grantham, R. and Enjily, V. *Multi-storey timber frame buildings: a design guide (BRE Report BR454)*. London: BRE Bookshop, 2003.

91   Structural Timber Association. *Gable wall spandrel panels (Advice Note 17)*. Alloa: STA, 2020.

92   UK Timber Frame Association. 'Timber Engineering Notebook series. No. 5: Timber frame structures — platform frame construction (part 3)'. *The Structural Engineer*, 91(7). July 2013, pp43–50.

93   Marcroft, J.P. *Disproportionate collapse of timber structures. Part 1: Literature review; Part 2: Permissible stresses in fasteners and behaviour of timber connections under short duration loading; Part 3: Full-scale testing programme simulating accidental events on a trussed rafter roofed building (TRADA Technology Research Report RR 3/93)*. High Wycombe: TRADA, 1993.

94   Salgo, M.N. 'Examples of timber structure failures'. *ASCE Transactions*, 121, 1956, pp588–600.

95   British Standards Institution. *BS 5268-6.1:1996: Structural use of timber — Code of practice for timber frame walls — Dwellings not exceeding seven storeys*. London: BSI, 1996.

96   CROSS UK. *CROSS Safety Report: The risk of collapse of multi-storey CLT buildings during a fire*. Available at: www.cross-safety.org/uk/safety-information/cross-safety-report/risk-collapse-multi-storey-clt-buildings-during-966 [Accessed: November 2023].

97   Fire Safety Wood in Construction. Available at: https://timberfiresafety.org/ [Accessed: November 2023].

98   CROSS UK. *Cross-laminated timber (CLT) in multi-storey buildings*. Available at: www.cross-safety.org/uk/safety-information/cross-feature-article/cross-laminated-timber-clt-multi-storey-buildings [Accessed: November 2023].

99   Law, A. and Hadden, R. 'We need to talk about timber: fire safety design in tall buildings'. *The Structural Engineer*, 98(3), March 2020, pp10–15. DOI: https://doi.org/10.56330/XJPS1661 [Accessed: November 2023].

100 Law, A. and Hadden, R.M. 'Burnout Means Burnout'. *SFPE Europe*, 5, Q1 2017. Available at: www.sfpe.org/publications/periodicals/sfpeeuropedigital/sfpeeurope5/issue5feature1 [Accessed: November 2023].

101 HM Government. *The Building Regulations 2010. Approved Document B: Fire Safety. Volume 1: Dwellings; Volume 2: Buildings other than dwellings*. Available at: www.gov.uk/government/publications/fire-safety-approved-document-b [Accessed: November 2023].

102　British Standards Institution. *BS EN 338:2016: Structural timber. Strength classes*. London: BSI, 2016.

103　pro:Holz. *Cross-laminated timber structural design: basic design and engineering principles according to Eurocode*. Vienna: pro:Holz Austria, 2014. Available at: www.proholz.at/fileadmin/proholz/media/shop_Publikationen/Information_pdf/cross_laminated_timber.pdf [Accessed: November 2023].

104　HM Government. *Party Wall etc. Act 1996*. Available at: www.legislation.gov.uk/ukpga/1996/40 [Accessed: November 2023].

105　Scottish Government. *Building standards technical handbook 2022: domestic*. Available at: www.gov.scot/publications/building-standards-technical-handbook-2022-domestic [Accessed: November 2023].

106　*CP 110-1:1972: Code of practice for the structural use of concrete — Design, materials and workmanship*. London: BSI, 1972.

107　Building Control Alliance. *The Building Regulations 2010 — England & Wales. Requirement A3 — Disproportionate collapse (BCA Technical Guidance Note 21)*. Available at: www.labc.co.uk/sites/default/files/resource_files/bca_guidance_note_21_disproportionate_collapse.pdf [Accessed: November 2023].

108　Matthews, S. *Structural appraisal of existing buildings, including for a material change of use. Part 1: Requirements for a structural appraisal; Part 2: Preparing for structural appraisal; Part 3: Structural appraisal procedures; Part 4: Additional considerations and information sources (BRE Digest 366 Parts 1–4)*. Watford: IHS BRE Press, 2012.

109　Bell, P. 'Party Wall etc. Act 1996: brief guidance for engineers'. *The Structural Engineer*, 94(2), February 2016, pp32–34. Available at: www.istructe.org/journal/volumes/volume-94-(2016)/issue-2/professional-guidance-party-wall-etc-act-1996-b/ [Accessed: November 2023].

110　Ostime, N. *RIBA job book*. 10th ed. London: RIBA Publishing, 2020.

111　British Standards Institution. *BS 5975:2019: Code of practice for temporary works procedures and the permissible stress design of falsework*. London: BSI, 2019.

112　The Institution of Structural Engineers. *Code of conduct*. Available at: www.istructe.org/about-us/istructe-code-of-conduct [Accessed: November 2023].

113　HSE. *Principles and guidelines to assist HSE in its judgements that duty-holders have reduced risk as low as reasonably practicable*. Available at: www.hse.gov.uk/enforce/expert/alarp1.htm [Accessed: November 2023].

114　HSE. *Assessing compliance with the law in individual cases and the use of good practice*. Available at: www.hse.gov.uk/enforce/expert/alarp2.htm [Accessed: November 2023].

115　HSE. *Policy and guidance on reducing risks as low as reasonably practicable in design*. Available at: www.hse.gov.uk/enforce/expert/alarp3.htm [Accessed: November 2023].

116　HSE. *HSE principles for Cost Benefit Analysis (CBA) in support of ALARP decisions*. Available at: www.hse.gov.uk/enforce/expert/alarpcba.htm [Accessed: November 2023].

117　HSE. *Reducing risks, protecting people: HSE's decision-making process*. Sudbury: HSE Books, 2001. Available at: www.hse.gov.uk/enforce/assets/docs/r2p2.pdf [Accessed: November 2023].

118　The Institution of Structural Engineers. *Appraisal of existing structures*. 3rd ed. London: IStructE Ltd, 2010.

119　*ISO 13822:2010: Bases for design of structures. Assessment of existing structures*. Geneva: ISO, 2010.

120　Construction Industry Council. *Definitions of inspections and surveys of buildings*. London: CIC, 2005.

121　CIRIA. *Structural renovation of traditional buildings (CIRIA Report 111)*. Amended ed. London: CIRIA, 1994.

122　BM TRADA. *Assessment and repair of structural timber (WIS 1-34)*. High Wycombe: BM TRADA, 2019.

123　Sutherland, J., Humm, D. and Chrimes, M. eds. *Historic concrete: background to appraisal*. London: ICE Publishing, 2001.

124　British Standards Institution. *BS 449:1932: The use of structural steel in building*. London: BSI, 1932.

125　LABC. *How to get it right: Removing a chimney the right way (with video showing the wrong way...)*. Available at: www.labc.co.uk/news/how-to-get-it-right-removing-chimney-right-way-video-showing-wrong-way [Accessed: November 2023].

126　The Concrete Society. *Design guidance for strengthening concrete structures using fibre composite materials (Technical Report 55)*. 3rd ed. Camberley: The Concrete Society, 2012.

127　Association of Rooftop & Airspace Development. *Promoting learning, innovation and good practice in airspace development*. Available at: www.arad.uk [Accessed: November 2023].

128　London District Surveyor's Association. *Guidance on achieving compliance on disproportionate collapse in existing buildings for Class 2B cases in single/multiple occupancy* [known as the Camden Ruling, internal document].

129　Greater London Council. *London Building (Constructional) Amending By-Laws 1970*. London: GLC, 1970.

130　Morton, J. *Accidental damage robustness and stability*. Rev ed. Windsor: Brick Development Association, 1996.

131　Heyman, J. *The Masonry arch*. Chichester: Ellis Horwood, 1982.

132 Bussell, M., Lazarus, D. and Ross, P. *Retention of masonry facades: best practice guide (CIRIA C579)*. London: CIRIA, 2003.

133 CROSS UK. *CROSS Safety Report: Scottish tenements*. Available at: www.cross-safety.org/uk/safety-information/cross-safety-report/scottish-tenements-26 [Accessed: November 2023].

134 CROSS UK. *CROSS Safety Report: Progressive collapse of an old mill building during a fire*. Available at: www.cross-safety.org/uk/safety-information/cross-safety-report/progressive-collapse-old-mill-building-during-122 [Accessed: November 2023].

135 British Standards Institution. *BS 9999:2017: Fire safety in the design, management and use of buildings. Code of practice*. London: BSI, 2017.

136 British Standards Institution. *BS 9991:2015: Fire safety in the design, management and use of residential buildings. Code of practice*. London: BSI, 2015.

137 British Standards Institution. *BS EN 1992-1-2:2004+A1:2019: Eurocode 2. Design of concrete structures — General rules. Structural fire design*. London: BSI, 2019.

138 British Standards Institution. *BS EN 1995-1-2:2004: Eurocode 5. Design of timber structures — General — Structural fire design*. London: BSI, 2004.

139 The Steel Construction Institute. *Structural fire engineering: Investigation of Broadgate Phase 8 fire (P113)*. Ascot: SCI, 1995. Available at: www.steelconstruction.info/images/2/2e/SCI_P113.pdf [Accessed: November 2023].

140 International Code Council and Society of Fire Protection Engineers. *Fire safety for very tall buildings: engineering guide*. 2nd ed. Cham: Springer, 2022.

141 The University of Edinburgh. *PIT Project: Behaviour of steel framed structures under fire conditions: main report*. Edinburgh: The University of Edinburgh, 2000. Available at: www.eng.cd.ac.uk/sites/eng.ed.ac.uk/files/attachments/research-projects/20150115/cardington-main-report.pdf [Accessed: November 2023].

142 Kirby, B.R. *Data on the Cardington fire tests*. [s.l.]: British Steel, 2000.

143 Bailey, C.G., Lennon, T. and Moore, D.B. 'The behaviour of full-scale steel framed buildings subject to compartment fires'. *The Structural Engineer*, 77(8), 20 April 1999, pp15–21.

144 Beitel, J. and Iwankiw, N. *Analysis of needs and existing capabilities for full-scale fire resistance testing (NIST GCR 02-843-1 (Revision))*. Gaithersburg, MD: NIST, 2008. Available at: https://nvlpubs.nist.gov/nistpubs/gcr/2008/gcr02-843-1.pdf [Accessed: November 2023].

145 The Roebling Construction Company. *Tests of the Roebling system of fire-proof construction*. New York City: The Roebling Company, 1899, p113.

146 Law, A. and Bisby, L. 'The rise and rise of fire resistance'. *Fire Safety Journal*, 116, September 2020. Available at: www.sciencedirect.com/science/article/pii/S0379711220304355 [Accessed: November 2023].

147 Spinardi, G., Law, A. and Bisby, L. 'Vive La Résistance? Standard fire testing, regulation, and the performance of safety'. *Science as Culture*, 2023. Available at: www.tandfonline.com/doi/full/10.1080/09505431.2023.2227186 [Accessed: November 2023].

148 International Standards Organization. *ISO 834-11:2014: Fire resistance tests — Elements of building construction — Part 11: Specific requirements for the assessment of fire protection to structural steel element*. Geneva: ISO, 2014.

149 British Standards Institution. *BS 476-20:1987: Fire tests on building materials and structures — Method for determination of the fire resistance of elements of construction (general principles)*. London: BSI, 1987.

150 Standards Australia. *AS 1530.4-2005: Methods for fire tests on building materials, components and structures. Part 4: Fire-resistance test of elements of construction*. Sydney: Standards Australia, 2005.

151 Underwriters Laboratories Inc. *UL Standard for safety: fire tests of building construction and materials (UL 263)*. 14th ed. Northbrook, IL: Underwriters Laboratories Inc., 2011.

152 Standards Council of Canada. *CAN/ULC-S101-14: Standard methods of fire endurance tests of building construction and materials*. 5th ed. Ottawa: Underwriters' Laboratories of Canada, 2014.

153 Lamont, S. *et al*. 'Assessment of the fire resistance test with respect to beams in real structures'. *Engineering Journal* (AISC), 40(2), 2003, pp63–75.

154 Usmani, A.S. *et al*. 'Fundamental principles of structural behaviour under thermal effects'. *Fire Safety Journal*, 36(8), November 2001, pp721–744.

155 Ashton, L.A. 'Fire and the protection of structures'. *The Structural Engineer*, 46(1), January 1968, pp5–12.

156 The Institution of Structural Engineers. 'Special Issue: Structural fire engineering'. *The Structural Engineer*, 96(1), January 2018. DOI: https://doi.org/10.56330/EUOP1461 [Accessed: November 2023].

157 Bisby, L. 'Structural fire safety when responding to the climate emergency'. *The Structural Engineer*, 99(2), February 2021, pp26–27. DOI: https://doi.org/10.56330/YLVY8201 [Accessed: November 2023].

158 Law, M. and Beever, P. 'Magic numbers and golden rules'. *Fire Technology*, 31(1), February 1995, pp77–83.

159 Flint, G. *et al*. 'Recent lessons learned in structural fire engineering for composite steel structures'. *Fire Technology*, 49(3), July 2013, pp767–792.

160 Cast Consultancy. *Modern Methods of Construction: Introducing the MMC Definition Framework*. Available at: www.cast-consultancy.com/wp-content/uploads/2021/03/MMC-I-Pad-base_GOVUK-FINAL_SECURE.pdf [Accessed: November 2023].

161 Meacham, B.J. 'Fire performance and regulatory considerations with modern methods of construction'. *Buildings & Cities*, 3(1), 2022, pp464–487. Available at: https://journal-buildingscities.org/articles/10.5334/bc.201 [Accessed: November 2023].

162 CROSS-UK. *CROSS Safety Report: Fire protection of light gauge steel frames (Report 1030)*. Available at: www.cross-safety.org/uk/safety-information/cross-safety-report/fire-protection-light-gauge-steel-frames-1030 [Accessed: November 2023].

163 CROSS-UK. *CROSS Safety Report: Fire protection to light gauge steel frame walls (Report 1116)*. Available at: www.cross-safety.org/uk/safety-information/cross-safety-report/fire-protection-light-gauge-steel-frame-walls-1116 [Accessed: November 2023].

164 CROSS-UK. *CROSS Safety Report: Volumetric modular buildings and fire (Report 1065)*. Available at: www.cross-safety.org/uk/safety-information/cross-safety-report/volumetric-modular-buildings-and-fire-1065 [Accessed: November 2023].

165 Buchanan, A. and Ostman, B. *Fire safe use of wood in buildings: global design guide*. Boca Raton, FL: CRC Press, 2022.

166 Kotsovinos, P. *et al*. 'Fire dynamics inside a large and open-plan compartment with exposed timber ceiling and columns: CodeRed #01'. *Fire and Materials*, 47(4), 2022, pp542–568.

167 Kotsovinos, P. *et al*. 'Impact of ventilation on the fire dynamics of an open-plan compartment with exposed timber ceiling and columns: CodeRed #02'. *Fire and Materials*, 47(4), 2022, pp569–596.

168 Kotsovinos, P. *et al*. 'The effectiveness of a water mist system in an open-plan compartment with an exposed timber ceiling: CodeRed #03'. *FPE Extra* (SFPE), 2022. Available at: www.sfpe.org/publications/fpemagazine/fpeextra/fpeextra2022/fpeextraissue74 [Accessed: November 2023].

169 Kotsovinos, P. *et al*. 'Impact of partial encapsulation on the fire dynamics of an open-plan compartment with exposed timber ceiling and columns: CodeRed #04'. *Fire and Materials*, 47(4), 2023, pp597–626.

170 Ronquillo, G., Hopkin, D. and Spearpoint, M. 'Review of large-scale fire tests on cross-laminated timber'. *Journal of Fire Sciences*, 39(5), September 2021, pp327–369.

171 Liu, J. and Fischer, E.C. 'Review of large-scale CLT compartment fire tests'. *Construction and Building Materials*, 318, 2022. DOI: https://doi.org/10.1016/j.conbuildmat.2021.126099 [Accessed: November 2023].

172 British Standards Institution. *BS EN 1993-1-2:2005: Eurocode 3. Design of steel structures — General rules — Structural fire design*. London: BSI, 2005.

173 British Standards Institution. *BS EN 1994-1-2:2005+A1:2014: Eurocode 4. Design of composite steel and concrete structures — General rules. Structural fire design*. London: BSI, 2014.

174 LaMalva, K.J. ed. *Structural fire engineering (ASCE Manuals and Reports on Engineering Practice no 138)*. Reston, VA: ASCE, 2018.

175 The Institution of Structural Engineers. *Introduction to structural fire engineering*. London: IStructE Ltd, 2020.

176 The Institution of Structural Engineers. *Guide to the advanced fire safety engineering of structures*. London: IStructE Ltd, 2007.

177 Buchanan, A.H. and Abu, A.K. *Structural design for fire safety*. 2nd ed. Hoboken, NJ: John Wiley & Sons, 2017.

178 LaMalva. K. and Hopkin, D. (eds). *International handbook of structural fire engineering*. Cham: Springer, 2021.

179 CIBSE. *Fire safety engineering (CIBSE Guide E)*. 4th ed. London: CIBSE, 2019.

180 Barnfield, J.R. and Porter, A.M. 'Historic buildings and fire: fire performance of cast-iron structural elements'. *The Structural Engineer*, 62A(12), December 1984, pp373–380.

181 Meacham, B. *et al*. "Fire and Collapse, Faculty of Architecture Building, Delft University of Technology: Data Collection and Preliminary Analysis," in *Proceedings, 8th International Conference on Performance-Based Codes and Fire Safety Design Methods*, Lund, Sweden, 2010.

182 Engelhardt, M.D. *et al*. 'Observations from the fire and collapse of the Faculty of Architecture Building, Delft University of Technology'. In *ASCE Structures Congress 2013: Bridging your passion with your profession*. Reston: VA: ASCE, 2013, pp1138–1149. Available at: www.researchgate.net/publication/271367709_Observations_from_the_Fire_and_Collapse_of_the_Faculty_of_Architecture_Building_Delft_University_of_Technology [Accessed: November 2023].